A STELLAR LIFE

TO ELAINE FOR HELP IN EDITING,

Helmut

A STELLAR LIFE

HELMUT A. ABT

KITT PEAK NATIONAL OBSERVATORY

TUCSON, ARIZONA

Charelston, SC
www.PalmettoPublishing.com

A Stellar Life

Copyright © 2020 by Helmut Abt

All rights reserved.

No portion of this book may be reproduced, stored in a retrieval system, or transmitted in any form by any means–electronic, mechanical, photocopy, recording, or other–except for brief quotations in printed reviews, without prior permission of the author.

First Edition

Hardcover ISBN: 978-1-64990-518-5
Paperback ISBN: 978-1-64990-604-5
eBook ISBN: 978-1-64990-602-1

Frontispiece

Helmut at age 45 years, fresh from Easter Island.

(Photo by John B. Irwin)

TABLE OF CONTENTS

1	Crater Elegante	1
2	The Crab Nebula Supernova	3
3	Introduction	10
4	High School and University in Illinois (1940-1948)	12
5	Caltech (1948-1952)	15
6	The Lick Observatory (1952-1953)	25
7	University of Chicago (1953-1959)	27
8	McDonald Observatory (1956-1959)	29
9	Kitt Peak National Observatory (1955-date)	32
10	Libraries	46
11	Gordon Grant	48
12	Spectral Atlases and Classifications	49
13	Visiting Professor Program (1960-1962)	54
14	Frequencies of Double Stars	57
15	Japan	61
16	The Unusual Scenery in Arizona and Utah	71
17	Pacific Islands	75
18	Marshall Space Flight Center (1967)	79
19	Easter Island (1970)	80
20	The Galapagos Islands (1971)	89
21	George Van Biesbroeck Prize	100
22	The Astrophysical Journal	102
23	Multiplicity of Solar-Type Stars (1976-2006)	116

24	Exoplanets	118
25	Catalogs	120
26	Publication Studies	122
27	China	134
28	Early Chinese Discoveries	151
29	Blue Stragglers	157
30	John Fountain	159
31	The Difference between Metal-Rich and Metal-Poor Stars	163
32	Local Interstellar Bubble	165
33	Spain	167
34	Jordan and Petra	170
35	George Ellery Hale	173
36	Thailand	175
37	Thailand, Laos, Malaysia (2004)	177
38	Culture	181
39	In My House	184
40	Health	189
41	Summary	192
	References	195
	Addenda - Stories about Astronomers	205
	Volume Index	231

1

CRATER ELEGANTE

One Christmas vacation Bill Miller and I decided to search for the remains of the Camino del Diablo (Highway of Death) in his jeep. That was the dirt road along the Arizona-Mexico border that some of the '49ers used to avoid the mountains farther north in going to the gold rush in California. There was little water along the road, only the natural spring at Quitobaquito Spring (Fig. 1-1) in what is now Organ Pipe National Monument and in the natural tanks in the Tinajas Altos Mts. After rains, water remained in the stone pockets in the latter, but one had to climb up to them. Skeletons farther down showed the people who did not have the strength to climb up. The Quitobaquito Spring has been in the news recently because of the evidence that this rare oasis in the Sonora Desert is now in danger of drying up due in part to the fence being built along the border. It cuts through the Tohono O'odham Reservation. The fence divides families, as did the Berlin Wall, and the passage of animals.

The *Arizona Highways* had just come out with an article about people who had found a way to climb down to the bottom of Crater Elegante (Fig. 1-2) in the Pinacate crater region, now called the Pinacate National Park in Mexico. We drove to that crater. Bill wondered if it was of meteoric or volcanic origin. Several months later he organized an expedition to it with Caltech Geologist Robert Sharp, USC meteor expert Frederick Leonard, Caltech student Barclay Kamb, Bill, and I. Barclay later was a geologist and Provost at Caltech and married Linus Pauling's daughter.

Fig. 1-1: The Quitobaquito Spring in Pipe Springs National Monument, a natural oasis in the Sonoran Desert just north of the Arizona-Mexico border.

The crater is 1.5 km across and 250 m (820') deep. We climbed to the bottom. Dr. Leonard found no meteoric material in or around it. Dr. Sharp called it a collapsed caldera, in which water seeping down encounters hot magna that produces a huge explosion, throwing rocks up to 19 km away. More recently that explosion has been dated as 32,000 years ago.

Fig. 1-2. Carter Elegante in Pinacate National Park in Mexico, just off the road to Rocky Point (Punta Penasco).

2

THE CRAB NEBULA SUPERNOVA

Fig. 2-1. A Hubble picture of the Crab Nebula. The red features are due to Hα radiation from the gaseous filaments. The central blue light is synchrotron radiation coming from the pulsar at the center.

The Crab Nebula (Messier 1) is a beautiful gaseous nebula (Fig. 2-1) in Taurus which is the result of a supernova that exploded in 1054 AD.

The Chinese astronomers kept records of "guest", or new, stars because they thought that they affected people, i.e. astrology. A Swedish astronomer (Lundmark 1921) collected 60 such records between 134 BC and 1828 AD. No. 36 in his list occurred in Taurus in 1054 AD and the Chinese position agreed with the position of the Crab Nebula.

Astronomers Lampland (1921) at the Lowell Obs. and Duncan (1921) at the Steward Obs. found that the Crab Nebula is expanding and Hubble (1928) suggested that the Crab Nebula is the remnant of that explosion. Duncan (Mayall & Oort 1942), Deutsch & Lavdovsky (1940) and Trimble (1968) derived explosion dates in the 12th century, but those are valid for only the radial motion of that turbulent nebula. The Chinese first saw the explosion on 5 July 1054 (Clark & Stephenson 1977). They said that it was visible in daylight for 26 days and at night for 653 days.

The Japanese (Duyvendak 1942) also recorded the bright star but their date in the summer of 1054 AD was uncertain.

In June 1953 Bill Miller and I explored the top of White Mesa for three weeks. It is a 180 m (600 ft.) high mesa (Fig. 2-2) on the Navajo Reservation in northern Arizona, northeast of Tuba City.

Fig. 2-2. In the background is White Mesa as seen from US160 at the bottom. Only Navahos are allowed to leave the highway; the Navaho police effectively keep pot-hunters off the reservation. (Photo by the author.)

He had obtained permission to explore the 25 km² top from the Navajo Council because it had never been explored by an archeologist, according to the Museum of Northern Arizona staff in Flagstaff. Potsherds told Bill that it had been inhabited in the 10-13th centuries, but not since. Among the thousands of pictographs he found one of a crescent moon next to a circle, the latter indicating something big or bright (Fig. 2-3). He knew that the Native Americans rarely showed anything astronomical. He wondered whether it was a drawing of an eclipse, which was not so rare as to cause them to draw it? Fred Hoyle (in residence in Pasadena at the time) suggested it might represent the Crab Nebula Supernova. With help from Walter Baade, Bill computed, using Brown's lunar tables, that on 5 July 1054 the new star was within 2 deg. of the Moon and it was in a crescent phase.

The next year Bill and I explored the lower part of Navajo Canyon on a governmental grant because it would be flooded by Lake Powell after Glen Canyon Dam was completed. In the upper part of the canyon I found a similar pictograph, but it was a reflection of the one on White Mesa (Fig. 2-4). Bill published (Miller 1955) a description of these.

Brandt & Williamson (1979) wrote to National Park Superintendents in the American southwest as to whether they had seen similar pictographs. Together with Mayer (1979) and Fountain (2000), 31 more were found, roughly divided between the ones in Figures 2-3 and 2-4.

Fig. 2-3. The Crab Nebula supernova pictograph that Bill Miller found on White Mesa. It is a drawing on the sandstone walls of a cave. It represents what one would see in the spring of 1054 AD. (Photo by the late Bill Miller)

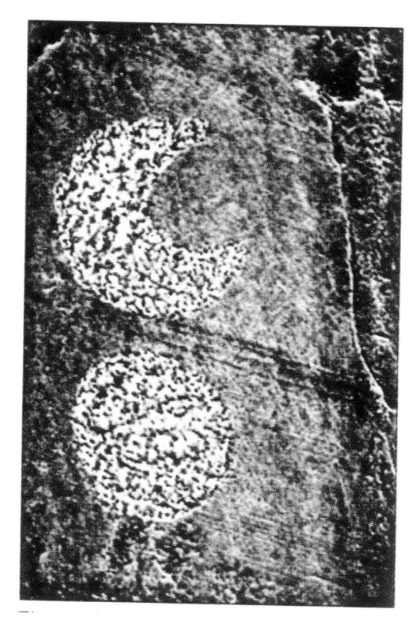

Fig. 2-4. The supernova pictograph I found in Navaho Canyon. It represents what one would see in the summer of 1054 AD. (Photo by the author.)

But was the supernova seen elsewhere? Collins et al. (1999) made a search of European and near eastern ecclesiastical records from the 11th and later centuries. They found seven accounts of a new bright star seen between 11 April and 20 May 1054 in the evening (and daylight!) sky before the Crab Nebula went into conjunction behind the Sun.

This leaves two dilemmas: why did the Asians not record the bright star in the spring before it went into conjunction behind the Sun and why did the Europeans not record the bright star in the summer after it came out of conjunction? We have only guesses, but they seem to make sense.

The Chinese astronomers undoubtedly saw the supernova in the spring, but in China at that time it was thought that an extremely bright guest star meant that the Emperor would die or that there would be a new dynasty. Telling the Emperor that he would probably die soon would be lethal for the astronomers, especially because the Emperor's favorite concubine had just died and he was in a depressed mood. The astronomers knew that the guest star would soon disappear behind the Sun, so maybe if they waited, it might not be there in the summer. They waited, but it was there in July and they had to report it. We do not know what happened to the astronomers, but the Emperor died the next year.

Why did the European church documents report the bright star in the spring of 1054 but not in the summer? That was the year when there was a serious effort to combine the Roman Catholic Church and the Eastern Orthodox Church. They worked on a plan for nearly a year, but announced on 16 July that they had failed. It is called the Eastern Schism. But there in the sky was a new bright star that could be seen in the daytime! Was that God's condemnation of their failed efforts? It might have been safer to the record keepers not to record the bright star in the summer.

However, the Native Americans evidently saw the supernova both in the spring and summer of 1054; that is why they drew it in the spring, like the Navajo Canyon pictograph, and in the summer, like the White Mesa pictograph.

There are still problems in understanding the type of supernova this was because none of the recent ones stay near maximum light for 80 days. We are learning that supernovae are more complicated and individual than we previously thought. Part of the above was published in Abt & Fountain (2018) in a Chinese publication.

3

INTRODUCTION

I am not writing this book because I think that people in the future will be interested in me. I never received any major awards or prizes and will be forgotten in several decades. I started some new areas of astronomy, but others later will repeat that work with better data and my early contributions will be forgotten. Instead, consider me as a wallflower who witnessed the rapid growth of astrophysics during the 20th century and knew many of the participants. I initiated new policies and technologies in publications, such as making the ApJ more international in content and participation, and converting to digital operation and publication. I will shock some people who still think that religious organizations should determine how people should live, as they did in 15th century Europe and where slavery was allowed. I visited places that later became overrun with tourists. I hope that you will find the following interesting after a few words of introduction.

My foster son, Daniel V. Conaway, suggested the title of this book and helped me reorganize it. Mike Peralta solved some major computing problems for me.

My family came to the US in 1927 after my father lost his furniture factory during the German inflation. I was two years old at that time. In the rapid inflation when the value of the money changed twice daily, people could not save up to bury furniture. Also laws prevented employers from laying off employees. My father set up as a furniture designer in the US, favoring the Bauhaus modern style while most Americams favored the colonial style. We lived in Jamestown, NY, the furniture capital of the country at the time.

Jamestown was the home of three famous people. One was Roger Tory Peterson (1908-1996), who published the first modern "Guide to the Birds", which sold out in one week. He published six later editions. There is a Roger Tory Peterson Institute in Jamestown.

The second famous person from Jamestown was Lucille Ball, a comedienne who became famous later for "I Love Lucy" on TV. There is a museum in Jamestown of her memorabilia.

The third famous person from Jamestown was Robert H. Jackson. He was a legal advisor to Franklin D. Roosevelt while the latter was Governor of NY. After Roosevelt became President, he appointed Jackson as General Counsel for the IRS, then US Solicitor General, then US Attorney General (1940-41), and then Associate Justice of the US Supreme Court (1941-1954). At the end of WWII Pres. Truman appointed him as the Chief Prosecutor at the Nuremberg trial of Nazi leaders. He returned to the Supreme Court in 1946 but died of heart problems in 1954. There is an active Robert H. Jackson Institute in Jamestown. In 1939 I worked after school in the massage facility in the YMCA. I changed sheets, cleaned up messes, ran errands, etc. When Jackson came in, my Hispanic boss whispered to me that he was a "famous lawyer". I remember Jackson as being thin, of moderate height, dark hair, and reserved.

My mother was an excellent pianist and I remember her playing the Schubert *Impromptus*. Her teacher in Jamestown was Mrs. Appisch, a pupil of Egan Petri, who was known for his beautiful touch. Petri was a pupil of Ferruccio Busoni and, in turn, Petri became the teacher of many famous pianists such Victor Borge, Eugene Istomin, Gunnar Johansen, Ozan Marsh, John Ogdon, and Earl Wild.

4

HIGH SCHOOL AND UNIVERSITY IN ILLINOIS (1940-1948)

As the furniture business in Jamestown gradually failed, my family moved to the Chicago area where my father worked for Storkline, a baby furniture company. Even in a depression new parents need furniture for a baby and are unlikely to settle for used furniture. My father died suddenly of a heart attack in 1940, leaving my mother, who had never worked for a living, to raise my brother (17) and me (15).

As a high-school student in Oak Park, IL and facing the WWII draft, I realized that I would just become a soldier unless I was in a university and qualified for a higher position. My brother Karl was a university student with language experience. In the army he helped interview captured German soldiers. They abided by international standards, asking only name, rank, and serial number. But some German soldiers were disgusted with the Nazis. His group learned that the Germans were amassing a million troops in the Netherlands for a last stand and it sent that information to the Allied headquarters. The info was ignored – hence the Battle of the Bulge where thousands of Allied troops were killed (Abt, K. 2004).

I arranged to pass my senior year courses by exam and entered Northwestern Univ. a half-year before my high school graduation. At

Northwestern I majored in math because I was unsure where I was headed, and soon obtained part-time employment by grading homework papers for math courses. Later I ran physics labs, obtained x-ray diffraction pictures for chemists, and tested aircraft rivets for an engineering professor. I failed the draft for a trivial reason – a pilonidal cyst, which had been cured. I soon shifted to physics and my M.S. thesis advisor was Dr. Russell A. Fisher.

Dr. Fisher had an interesting time during WWII. He was a member of the ALSOS Mission. The US wondered whether the Germans were working on an atomic bomb. The mission consisted of military, scientific, and intelligence personnel. The head was Brigadier General Leslie Groves ("alsos" is the Greek word for "grove"), who later wrote a book about the mission. Dr. Samuel Goudsmit was the Chief Scientific Advisor. He had not worked for the Manhattan Project, so if captured by the Germans, he could not be forced to reveal any American secrets. He had 33 scientists working with him. One was Dr. Russell Fisher.

Goudsmit had been on the University of Michigan faculty, where he and George E. Uhlenbeck in 1925 proposed that electrons have spins. He was Jewish and born in the Netherlands. His parents were killed by the Nazis in 1943. Later Goudsmit headed the Physics Dept. at NU (1948-70), was at the Brookhaven National Lab. (1952-60), and was Editor-in-Chief of the *Physical Review* (1950-74). He started the *Physical Review Letters*.

The ALSOS scientists reasoned that if the Germans were working on an atomic bomb anywhere is western Germany, some radioactivity would get into the Rhine River water. So during the Battle of the Bulge, they had Fisher crawl half way across a bridge over the Rhine under bombardment and lower bottles to collect samples of the water. They had no way to test the water, so they sent the three bottles to Washington, DC for testing. Because the bottles would rattle in the box they had, they added a bottle of French wine with a joking label "Test this for radioactivity too". The reply came back "No radioactivity in the Rhine River water, but there is in the French wine. Send more wine!" The ALSOS scientists felt that the people in Washington just wanted French wine, and ignored the request. But the request went up the official chain until they could not ignore it. In some places the soil is naturally radioactive, so they sent Fisher to southern France from where the wine came. Fisher collected samples of the soils, water, grapes, and wines, and sent them all to Washington. No further requests.

Although physics, especially spectroscopy, was fun, a course in astrophysics taught by Wasley S. Krogdahl using Goldberg & Aller's book really turned me on. I received admittance from Princeton but they specialized in theoretical studies. The University of Chicago also would accept me but had no available student financial support. However I learned at Yerkes Observatory from Drs. Struve, Kuiper, and Greenstein that Caltech was going to start an Astronomy Dept. in the fall of 1948, just after the Palomar 200-inch (5.1m) was dedicated. I applied and received acceptance and an assistantship.

5
CALTECH (1948-1952)

In August 1948 I took the Sante Fe train from Chicago to Pasadena (38 hours). I had shipped my bike ahead to the Pasadena YMCA and arranged to stay there. My Assistantship paid $100 per month and I doubted that I could live on that amount. I went to the Caltech housing office to find out what housing was available. It had an offer from a local doctor and his wife (Dr. John and Virginia Bolenbaugh) for a free room to a Caltech student in exchange for baby-sitting for their two boys, Andy (6) and Rich (4), three or four nights a week. Their house (3575 San Pasqual St.) was at the east end of Pasadena, three miles from Caltech, but that was OK on a bike because it seldom rained there.

I rode there and met Virginia and the boys, and we liked each other. However Virginia said that I should first get the approval from her husband at his office on Hill St. He agreed, but was amused that Virginia had already given me a house key!

Jack and Virginia were much in love with each other, although they were very different. Jack was thin, had a great sense of humor, and loved company. He loved his liquor, especially after Virginia died of a brain tumor. He probably was not a very well read doctor, but his bedside manner was outstanding. I suspect that many of his patients went to him because he knew how to make them feel better. He also consulted for a mental hospital in Whittier.

Virginia was taller, plump, and very practical. I never met a person, male or female, who knew their way around the streets of Los Angles better. Any time a new store opened anywhere in LA, she and her friend Helen Drummond drove there to see it. She was an excellent cook. I loved her. Even after I left Caltech, Jack and Virginia came to Tucson or I went there several times a year.

Jack and Virginia were very good to me. I ate my breakfasts and half of my dinners with them. Lunches were with Allan at the Greasy Spoon on the campus. The B's took me on day trips. My first trip to Mt. Wilson Observatory was with them. I was shocked and frightened at the steepness of the San Gabriel Mts. They also took me to Santa Barbara to see the Franciscan Cathedral originally built in 1786, but after a fire the current building was rebuilt in 1815-20. We went to Ensenada, Mexico, where later the *Titanic* was filmed. We went to Long Beach once where I saw my first Flamenco dancer, which really turned me on. We also went to plays in Hollywood, where I remember seeing Charles Laughlin and Maureen O'Hare. I had shipped to Pasadena my hand-made record player and introduced the B's to some of their first classical music, like Stravinsky's *L'Histoire du Soldat* and *Octet*, and Strauss's *Also Sprach Zarathustra*. They didn't care for the old classics like Mozart and Beethoven.

Jack once invited me to his office to watch while he circumcised a baby. It was a one-minute procedure. After cleaning the penis, he pulled the foreskin forward with one hand and sniped it off with the other. There was no bleeding and the baby did not cry (I understand that now, as a precaution, a local anesthetic is used). It was like trimming a fingernail. I asked "Aren't you going to sew the two edges of the shaft skin together?" and he said that that was not necessary; they would find each other and grow together.

The operation is much more complicated when done to an adult, taking nearly an hour. Some people say that that procedure should wait until the child becomes an adult and can decide for himself whether he wants it. (Should parents also subject their children to no religious training until they are adults and can decide for themselves?) The response is that parents make many decisions for their children (such as names, vaccinations, diets, behavior, religion) for their children's health and futures. Removing the foreskins, which harbor viruses, helps prevent venereal diseases (e.g. AIDS) by a factor of approximately six in adults. It became routine for babies in the US after thousands of soldiers returned from Europe after WWI and were found to have syphilis (for which there was no cure in the 1920s and 1930s). Those were primarily the soldiers who were not circumcised. The UN recommends that in regions of AIDS epidemics all men and boys be circumcised as the cheapest and most effective way to stop the epidemics. There is currently an AIDS epidemic in Beijing, so the government has recommended that all university male staff and students, aged 17-24, be circumcised and thousands have been.

Rich and I remained excellent friends for the 68 or more years after I left Pasadena. People attribute his exaggerated sense of humor to my influence. His first wife, who gave him two daughters, left him but I love his second wife, Pamela. After Rich's parents died, we met several times a year and they helped me, especially after John died. Rich worked for the LA Times and Pam for the phone company. They lived in Houston for a while but then returned to inherit their parent's house on San Pasqual St. They kept it in excellent condition; they installed a pool and a wonderful collection of plants.

The new Astronomy Dept. at Caltech had Dr. Jesse S. Greenstein as Chair and Drs. Guido Munch and Albert G. Wilson as faculty. There were four students in the first class: Allan R. Sandage, Morton S. Roberts, James Parker, and me. Mort quit at the end of the first year because it was too hard. He taught in a small college for a couple years and then got a Ph. D. from UC Berkeley. He spent most of his career at the National Radio Astronomy Obs. and was its Director part of that time. Jim Parker also quit at the end of the first year.

Allan and I became best friends and helped each out with classwork. We ate many meals together and appreciated the Met operas on the radio. In the last phone conversation that I had with him before he died on 13 Nov. 2010 of pancreatic cancer, I told him why he was right in his dispute with Gerard de Vaucouleurs about the lack of absorption toward the north galactic pole. It was because I found (Abt 2011) that an unknown supernova about 160 million years ago swept virtually all the interstellar gas out of the region within about 150 pc of the Sun, so there was no appreciable absorption toward the galactic pole. I miss Allan.

The astronomy students were required to take several of the basic physics courses for all physics graduate students. One was optics under Dr. Robert King. Those courses were primarily problem courses, so during the first class Dr. King asked us to do the first five problems for the next Monday. I worked on them and got nowhere. Allan did the same. We worked together for days and got nowhere. Next Monday Dr. King asked Mr. Farrell if he had the first problem. Richard Ferrell went to the blackboard and in two lines he had the answer. Dr. King asked Eugene Parker if he had the second problem. Parker went to the board and in two lines had the answer. We learned later that Ferrell was the only person to go through undergraduate work at Caltech with a straight A average. He was also the only Caltech student to be able to do all the problems in William Smythe's *Static and Dynamic Electricity*. Eugene Parker later explained the origin of the solar wind and in recognition for that the *Parker Solar Probe* spacecraft was named for him.

Allan and I were devastated. We were convinced that we would flunk out of Caltech. We agreed that we could not return to our homes in disgrace, but decided that we would get jobs in California, perhaps as milkmen. But we decided that we would not quit; we would stay until they kicked us out. They never got around to doing that, so we graduated. For a person who never got a C in a course before, I was sometimes glad to get Cs. Years later I met a fellow classmate who said "You and Sandage seemed to know what was going on. The rest of us students sat in the back row and didn't have a clue." So I must have had a lot of people fooled.

It was that experience of having to work hard all the time that left me with a lifetime drive to work and develop considerable humiliation. Once a month Allan and I went to a movie and then felt guilty for taking two hours off of work.

Greenstein augmented the courses by himself, Munch, and Wilson by bringing in the best astronomers to teach us one-quarter courses, namely Jan Oort, Martin Schwarzschild, Bengt Stromgren, and others. In his last year Allan took off to go to Princeton for a half year to work with Schwarzschild. They produced the first stellar interior model for stars leaving the main sequence. Both Allan and I did individual research projects in our summers. He did a curve-of-growth for solar vanadium lines and I detected hyperfine structure in the solar spectrum. In fact, I would observe on Mt. Wilson using the 150-foot (46 m) solar tower in the afternoons and observed Wolf-Rayet stars in the nights with the 60-inch (1.5 m) reflector.

Several times I walked down the ridge below the Monastery (the name of the buildings where the astronomers lived and ate) to the small laboratory building where Albert A. Michelson measured the speed of light in 1924-1926. He sent a light signal to a rapidly-rotating mirror on Mount San Antonio (Baldy) 22 miles away and recorded the returning signal. The unlocked laboratory looked like he had just walked away from it the day before with all the tools and parts lying around. In 1907 he became the first American to win the Nobel Prize for his part in the Michelson-Morley experiment. I learned that that historical lab survived the fires on Mt. Wilson in 2019 and 2020, but I have worried about it being vandalized by tourists.

I enjoyed working for Dr. Olin C. Wilson 20 hours per week for my assistantship, riding my bike up Lake St. to 813 Santa Barbara St., the headquarters buildings of the Mt. Wilson Observatory. Olin had obtained a Ph.D. at Caltech in physics before it had an Astronomy Dept. He was short, plump, and always wore a shapeless sport coat with overstuffed pockets. He had a mustache and beard. One morning he shaved off the beard but at breakfast his wife and two children did not notice the difference. He was a laid-back pipe-smoking capable patient astronomer. He was politically liberal and financially conservative. After his death I officiated at his memorial in the Caltech Athenaeum and I wrote his *Biographical Memoir*. But he and other Mt. Wilson & Palomar astronomers suffered from its policy of denying observing time on telescopes after the age of 65. Supposedly that was to give younger people the telescope time. For instance, just before reaching 65 Walter Baade discovered the two population types in M31 but could not follow the implications of that. He returned to Germany and died two years later, probably depressed. What a loss because of a stupid rule!

The one who survived that system was Alfred H. Joy (1882-1973), who kept a small office during retirement and did research on plates in the files until his 90s. I visited him often after moving to Tucson on my monthly trips to Pasadena to collect data on Mt. Wilson radial velocities. He always greeted me with smiles and enthusiasm. When I asked once what he was doing, he said that he was classifying M dwarfs from plates taken for radial velocities. I said "Great! That's an important need. Where are you going to publish the results?" (I was Editor of the ApJ then.) He said that the photoelectric astronomers didn't think that such types were worth publishing. I said "Bunk! They can't differentiate between dM and dMe stars". I said that if he sent me the material, I would see it through publication. Several years later I got the third or fourth carbon copy of his typewritten table. I worked over the material, adding statistics and results, and sent the text to him. He just had time to add some corrections before he died on 18 April 1973. It was published as Joy & Abt (1974) and has been cited hundreds of times.

Olin got an extension of five years to observe and measure calcium K-line strengths, indicating solar-like cycles in other stars. But when a new freeway in Pasadena destroyed his and Katherine's house, they moved to their daughter's city of West Lafayette, IN where their daughter was married to a chemist at Purdue Univ. I visited them once and sensed that Olin was sad to be away from astronomy, despite splurging on having a billiard table. We played a few games of Cowboy. Observational astronomers used to play billiards (a game called Cowboy) on cloudy nights when they couldn't observe; now they watch TV.

But Olin was another astronomer who had to give up research because of an arbitrary age limit. I am glad to see that others (Chris Corbally and Richard Gray) are continuing Olin's work by looking for solar cycles in young solar-type stars.

Olin Wilson and I did one small paper together on an Of binary; we got the wrong result because we did not use a high-enough dispersion. But for most of the four years I worked with him on spectra of zeta Aurigae.

Zeta Aurigae is one of those situations just made to yield much information about the outer atmosphere of a bright giant star. It consists of a B8 V star in an edge-on eclipsing system with a K4 II star and having a period of 2.66 years. As the small B star passes behind the outer atmosphere of the giant K star, the light from the B star can be seen for eight days, corresponding to 54 million km in the atmosphere, before it disappears. In the blue-violet region only the Balmer lines from the B star can be seen with superimposed lines from light passing through the K star atmosphere. In 1947-1948 Olin obtained spectra daily with the 100-inch (5.5 m) coude spectrograph, both going into eclipse and coming out. It is on those spectra that I worked for four years.

The spectra yielded temperatures, pressures, densities, and turbulent velocities at eight heights in the atmosphere. The surprising result was that because the atoms in the lines of sight were not full ionized by the ultraviolet light from the B star, the material was not smoothly distributed, but must exist in small dense clumps. That had been totally unexpected. It was published as Wilson & Abt (1954), which was the first *Astrophysical Journal Supplement* published. The second *Supplement* was Alfred Joy's paper on Mira Ceti and the third was my thesis.

In the summer of 1949, coming back from Chicago, I got off the Sante Fe in Williams and took the spur train to the Grand Canyon. I checked my suitcase at the train station and took my backpack, sleeping bag, and 2-man mountain tent with me. I hiked along the canyon rim to the head of the Kaibab Trail. After camping during the night, I hiked down (7 miles) and stayed at the campground. I saw no one on the trail down, at the campground, or on the Bright Angel Trail coming out, except for the mule trains taking supplies down. There were people at the Phantom Ranch at the bottom, but I guessed that meals and lodging there would be too expensive for me. I swam in Bright Angel Creek, watched the swiftly-moving Colorado River, and loved the fantastic scenery while hiking down and at the bottom. Many times in later years I made that hike, twice hiked (three days) across to the N rim, and twice to Clear Creek (18 miles each way). But never after 1949 did I feel that I was the only person hiking in the canyon.

In the summers at Caltech I worked on hyperfine structure in the solar spectrum and my thesis. I couldn't resist riding my bike during the lunch hours to the Bolenbaugh's and taking Andy and Rich across the street to Margie's pool.

It was in those years at Caltech that I had the chance to visit parts of the Southwest, such as the Grand Canyon, Bryce, Zion, and Cedar Breaks National Parks (during a one-week trip with Steve and Jack, who ran the mental hospital where Jack Bolenbaugh consulted). Soon I was taking trips during Christmas and summer vacations with William C. "Bill" Miller in his Jeep. Bill was the photographer employed by the Mt. Wilson & Palomar Observatories after the 200-inch was completed. His job was to get the best photographs with it. He cooperated with Eastman Kodak in Rochester, NY, which was endeavoring to develop faster films and extended sensitivities into the red. They sent their experimental emulsions to Bill and he tested them in his lab and on the telescope with long exposures. He also learned how to correct for the reciprocity-law failure for different colors. There were many trips with Bill until he died of a brain tumor; I described two in Sections 1 and 2.

When it came time to select a thesis topic, Dr. Greenstein had two suggestions. I selected the one that involved the variable star W Virginis. W Vir is a Population II Cepheid with a period of 17.27 days. Such stars generally occur in globular clusters or at high galactic latitudes; they have low metallicities, peculiar light curves, and slightly variable periods. Sanford (1949) found that it had double lines at times. Why?

I obtained 18 photographic spectra with the Mt. Wilson 100-inch (4.0 m) coude spectrograph to augment the five spectra obtained by Sanford. Each exposure was an all-night exposure. For 20% of the cycle there were double lines in which faint shortward lines appeared and grew in strength while the longward lines gradually disappeared. I obtained abundances, temperatures, pressures, opacities, and radial motions for the two sets of lines. What was happening apparently was that the pulsation wave from below the atmosphere had turned into a shock wave with Mach 5. The longward lines came from the in-falling part of the atmosphere and the shortward lines came from the region moving outward from behind the shock front. There were hydrogen emission lines from the shock front, which Kovtyukh et al. (2011) found was only one meter thick.

Not knowing much about shock waves, I visited Dr. Louis G. Henyey in Berkeley for info. He insisted that a compression factor of more than Mach 4 was inconsistent with the Rankine-Hugoniot shock wave equations and that this could not be a case of shock waves moving through a low-density atmosphere. Upon his (incorrect) advice, I mumbled in the final paper about inhomogeneous shells passing through one another. The correct analysis involving shock waves was done by Kovtyukh et al. (2011).

Because Allan Sandage was delayed a year by going to Princeton to work with Schwarzschild, I was the first person to get a Ph.D. in astronomy at Caltech and the only astronomy student graduating (cum laude) in 1952. I tell people that I graduated at the bottom of my class of one.

I bought a 1942 used Jeep and Bill Miller taught me how to drive. I received a Ph.D. before I got my first driver's license.

6

THE LICK OBSERVATORY (1952-1953)

Surprisingly, I never applied for a position in my life, even for childhood jobs. In those days, a potential employer phoned the head of another organization and asked whether he or she had any good students or employees. The new position was all arranged behind his/her back. So Director Charles D. Shane of the University of California's Lick Observatory phoned Jesse Greenstein, who recommended me. The position was meant to be a post-doc, although after I left it, it became a staff position for Dr. William P. Bidelman. I went there in June 1952.

At that time the astronomers lived on the mountain top, where there was a one-room school. Supplies and mail were brought up daily. An eight-room dormitory served the single people (like Olin Eggen, Pat Roemer, and I) while families (Jeffer's, Kron's, Mayall's, Shane's, and Virtanen's) had individual houses. The Shane 120-inch (3 m) reflector was under construction with a thin primary that had been cast to be a test flat for the Palomar 5-m. The only working telescopes were the 36-inch refractor, 36-inch Crossley reflector and an astrograph. Years later the astronomers and their families moved to a new UC Astronomy Dept. in Santa Cruz. I lived in the dormitory with a rented refrigerator to keep ice cream and cokes. It has been said that money cannot buy happiness, but it can buy ice cream! I enjoyed the time to read books and take trips to the beach, San Francisco and Pasadena, although I continued to be brain-washed into spending most of my time working.

The only high-dispersion spectrograph was the Mills spectrograph on the 36-inch refractor, giving 11 A mm^{-1} in a three-prism configuration. I decided to use it to see if RV Tauri stars had double absorption lines like W Vir. RV Tauri stars were also Population II variables but with longer periods than W Vir stars. I obtained spectra of U Mon (45^d), R Scu (73^d), and AC Her (37^d), each for more than a half year. Each showed double absorption lines at times. Only the work on U Mon led to a paper (Abt 1955), which has generally been ignored by others.

The reason why Pop II stars have shock waves in their atmospheres while Pop I Cepheids do not is because Pop II originate from main-sequence stars of one solar mass while those of Pop. I come from B stars of 10 solar masses; therefore, the surface gravities and speed of sound are less in the former, relative to the pulsational wave speed.

7

UNIVERSITY OF CHICAGO
(1953-1959)

The next offer came from Dr. Aden B. Meinel (1922-2011) at the Yerkes Observatory. He was a brilliant astronomer (IQ 182) whose career can fill hundreds of papers. Fortunately, his biography is being written by James B. Breckinridge and Alec M. Pridgeon, who have a contract with the Oxford Univ. Press to finish it by Dec. 2020, so I need not review his whole biography. Briefly, Meinel started the Kitt Peak National Observatory, the Multi-Mirror Telescope, the Optics Center at the UA, the LAMOST telescope in China, and secret projects for NASA and DOE in Nevada. He offered me a position on his Auroral Project, which already employed Joseph W. Chamberlain and Charles Fan. However, I had little interest in aurorae and felt incapable of designing a spectrograph to study them. That is the only time in my life that I failed to do what was expected, but Aden had many "irons in the fire" and we worked on other things.

Aden sat in on William W. "Bill" Morgan's course on the spectral classification of stars and immediately thought of ways in which he could help. One way was to design and build an 8-inch f/1 Schmidt camera with a field of 12°. It was sent to the McDonald Observatory and strapped onto a small telescope there. Exposures through a red filter with a half-width of 270 Å centered on Hα took 30 min. and through with an Hα interference filter took four hours. The published paper (Abt et al. 1957) concentrated on the Gum Nebula.

The Gum Nebula was named for its Australian discoverer, Colin Gum (1952). It is in the Vela-Puppis region, has a size of 15°x22°, has a distance of about 250 pc, namely outside of the Local Interstellar Bubble, and is probably excited by stars γ^2 Vel (WC7) and ζ Pup (O5f). No camera at that time could photograph the whole emission nebula, so a dozen individual frames and a drawing were published.

After the publication of the study of the Gum Nebula, Aden and I became involved in the search for the best site for the National Astronomical Observatory (see Section 9), which ended up on Kitt Peak. I spent parts of 1955 and 1956 on the site survey. While the proposed sites were being tested, I returned to the Yerkes Obs. until 1959. I became an Assistant Professor at the Univ. of Chicago and did research projects based on spectra from the McDonald Obs. My students included John C. Brandt, Arnold Heiser, Eugenio Mendoza, Elliot Moore, Peter Pesch, Carl Sagan, and Clayton Smith. Several of those became observatory directors, and outstanding research and teachers. It became apparent to me that each of these students was different. For instance, Peter Pesch had deep and broad interests in astronomy, not just in solving small individual problems. Later as a professor he tried to get his students to look at the larger picture and question physical ideas. Carl Sagan too had broad interests in science. He became well-known for explaining and interpreting astronomy on TV and in lectures. He learned how to convey his enthusiasm with simple language about our knowledge and ideas about astronomy to the general public. He got many people with no knowledge of astronomy to kindle their interests in the universe outside the earth. I too felt that I could do that, for instance in Visiting Professor Program talks and courses I gave for the Learning Curve, courses for seniors in Tucson who simply wanted to learn.

8

MCDONALD OBSERVATORY

(1956-1959)

The McDonald Observatory started when a Texas banker left $1 million in his will to the Univ. of Texas to build an observatory. He had no close relatives, but when the bequest became known, four distant cousins appeared and contested the will in court. It was during the Depression and to give $1M to astronomy seemed ludicrous to them. A court case occurred and Dr. Joel Stebbins of the Univ. of Wisconsin was asked to defend McDonald's bequeath to a jury of ranchers and farmers. Joel talked about the glories of astronomy and its philosophical importance. He saw that his defense was going over the heads of the jurors. So finally, he said "Perhaps if we learned more about astronomy, we could predict the weather better." That was a winning statement during those times of the drought and the dust bowl.

So the Univ. of Texas won the case and received $800,000, but the court gave $50,000 to each of the contesting four relatives. During the Depression that seemed like a huge amount of money, so each of them quit their jobs, bought cars and houses, and soon they were broke. At 2% interest they could have lived on their $50,000 but not to buy houses. It ruined their lives.

The Univ. of Texas had one astronomer and did not have the staff and facilities to build an observatory, so it signed a contract with the Univ. of Chicago, one of the leading astronomy groups in the country, to build an observatory. Dr. George Van Biesbroeck of Yerkes searched for the best site in Texas and selected Mt. Locke in W Texas. An 82-inch (2.1 m) telescope was built by the Warner & Swasey Co. in Cleveland and dedicated in 1939. Astronomer W. Albert Hiltner and others designed the auxiliary equipment, such as the excellent coude spectrograph.

The McDonald coude spectrograph was a joy to use and gave exciting results. Observing runs for Yerkes staff members were generally two weeks long, one for dark-sky observing and one for spectra of stars. In 1956 I was offered a four-week run and that was an opportunity to study the variability of supergiants. The supergiant region in a color-magnitude diagram is bisected by the Cepheid strip, a region of stars with very-periodic radial pulsations. But what about the stars to the left and right, e.g. Rigel (B8 Ia) and Deneb (A2 Ia)? Were they also variables?

Fortunately, all of the nights in the four-week run were clear and I obtained spectra of nine supergiants each night. I made preliminary measurements of the spectra in the afternoons to check whether something was happening in them. It turned out that all supergiants are pulsators, with those in the Cepheid strip having single periods and the supergiants to the left and right having multiple periods (Abt 1957).

Supergiant stars have very narrow spectral lines, showing little broadening due to rotation, whereas giants and dwarfs show large variations from star to star. What are the sources of line broadening in supergiants? I found that among bright giants (class II) it was rotational broadening with a mean V sin i of 26 km s^{-1}, consistent with the rotational velocities of the B stars from which they originated. They do not rotate like rigid bodies. Among the less-luminous supergiants (class Ib) there was a small amount of rotation but mostly macro-turbulence with the values consistent with those derived from curves of growth.

Not only was McDonald Obs. an excellent place for observing, but I liked the area. In the afternoons I climbed most of the peaks around there. A favorite place to go in the afternoons was Balmorhea State Park. It is a natural crystal-clear spring-fed swimming pool, 1.3 acres is area, 25 ft. deep with water that is always at 72-76 deg., all year around. The staff and astronomers at McDonald had picnics at the Rock Pile, a naturally formed pile of rocks about 100 ft (30 m) high through which one can walk and climb. Hispanic staff members cooked chickens and pigs in the Mexican way – buried underground with heated rocks.

9

KITT PEAK NATIONAL OBSERVATORY (1955-DATE)

In the 1950s the major observatories were in California (Lick, Mt. Wilson, Palomar) because its weather is good for observing. Universities in other parts of the country had small telescopes outside of the metropolitan areas, but they were usually in areas with poor weather for observing. Occasionally eastern astronomers went to California on sabbaticals or extended stays, such as Carl Seyfert. He worked for Rudolph Minkowski who had found some peculiar galaxies that were later called Seyfert Galaxies. Train trips to California took three to four days each way and astronomers in small departments could generally not be away for several weeks. However, in the 1950s the US developed a system of commercial flights (with DC-3s) that allowed a person to fly from any major city to any other in the country in one day. That meant that an astronomer could fly to an observatory in a good climate, observe for several days, and return, missing only a week of teaching.

At a conference in Flagstaff in 1954 of astronomers using photoelectric equipment, John Irwin (Indiana Univ.) proposed that the US should build a national observatory for all astronomers to use and located in the best place for observing in the continental part of the country. Dr. Robert R. McMath (Univ. of Michigan) took the initiative to obtain a grant from the National Science Foundation (NSF) to find the best place in the continental US for a national observatory. It was suspected that Hawaii and some foreign locations might be better but travel there for many astronomers was thought to be prohibitively expensive. McMath selected Aden B. Meinel (Yerkes Observatory, Univ. of Chicago) to head the project because of his proven ability with designing equipment and studying atmospheric conditions. He and I were both at Yerkes and I was known to be somewhat acquainted with the American southwest.

Two important requirements (there are many lesser ones) for an observatory are good clear weather and excellent "seeing". By the latter we mean sharp star images and high resolution images of extended sources, such as planets, nebulae, and galaxies. When the sunlight heats the ground anywhere during the day, that ground gives off heat waves at night, like what one sees above a house radiator below a window. Only for a sharp mountain peak where the heated surrounding ground is far below does one get good seeing. Also, that peak should have vegetation on the sides to avoid the sides producing heat waves. That leaves out flat Florida and leads to a mountain in the US southwest that is not barren. I had aeronautical maps of the southwest and knew the locations, elevations, and sizes of the mountains there. We avoided sites above 8,000 or 10,000 ft. on grounds that astronomers from low elevations might have trouble acclimatizing to higher elevations – a factor that is has proven to be important for 14,000-foot Mauna Kea Observatory in Hawaii.

Our first step was to hire a pilot. Marlin Krebs, superintendent of the McDonald Observatory, recommended John O. Casparis from Marfa, TX who had had logged 33,000 hours of flying experience. The airport in Alpine, TX was later named the Alpine-Casparis Municipal Airport. We arranged for him to fly me in his one-engine Cessna-140 plane around all the proposed mountain sites in western Texas and southern New Mexico and Arizona, starting on May 7, 1955. We had already decided to eliminate California, which has excellent summer weather and less good winter weather whereas Arizona and New Mexico have better winter weather but a cloudy monsoon season in July and August. In other words, a new major observatory should be placed in a complementary climate. I did look (with Leon Salanave) at Junipera Serra Peak, a cone on the coast near King City. However, we foresaw that the avid environmentalists would object to us building on that pristine peak.

Casparis flew me around sites in western Texas and southern New Mexico and Arizona for three days. These are listed in Table 1 in the order in which we flew. This list has not been published elsewhere. We flew around each peak several times while I took notes and pictures with my 35-mm camera, but with the barf bag handy.

TABLE 1: MOUNTAINS EXPLORED IN MAY 1955

MT.	LOCATION	ELEVATION	COMMENTS
Franklin	N of El Paso	7192'	Sharp, craggy, too close to El Paso.
Potrilla	W of El Paso	5957	Three conical peaks; W Mts., Dark volcanic rocks.
Florida	SE of Deming	7448	Rugged, barren
Cedar	SW of Deming	6215	Barren
Little Hatchet	SE of Lordsburg	6678	Barren

Big Hatchett	N Animas. On Continental Divide. Snow above 8000'	8366	Fairy rugged, some trees on it
Chiricahua	S of San Simon	9759	Forested but too high? Smoke from Douglas smelter in the valleys.
Pendregosa	S of Chiricahua Mts	6510	Steep cliffs on N; rugged. Smoke from Douglas.
Miller Pk.	Huachuca Mr.	9466	Too high? Huachuca Mts. Probably too close to Sierra Vista.
Apache Pk	Whetstone Mts	7711	Little room, too close to Ft. Huachuca? Smelter smoke?
Mt. Hopkins	Santa Rita Mts	8553	Forested, some toom but too close to Tucson.
Tumacacori		6440	Barren, rugged.
Baboquivari		7734	Too barren and steep
Coyote, Quinlan Range		6529	Barren, rocky, steep, and little room on top.
Kitt Peak		6875	Room, some trees, no road to the top! Investigate!
Santa Catalina		9157	Room & vegetation, but too close to Tucson lights
Rincon		8664	Similar to Catalinas.

Dragoon Mts.	E of Benson	7520	Vertical slabs, little flat space.
Swisshelm	N of Douglas	7185	These & Chiricahua ruled out by smelter smoke from Douglas.
Winchester	S end of Galiuro Mts	7631	Vertical slabs, no flat area, few trees.
Bassett Pk.	Galiuro Mts	7663	Like the Winchester Mts
Kennedy Pk		7116	
Signal Pk	S of Globe. Pinal Mts	7812	Road to top, heavily forested.
Aztec Mt.	NE of Roosevelt Lake in the Sierra Ancha Mts. in the Tonto Nat. Forest.	7694	Heavily forested, lots of room on top.
Four Peaks	NE of Phoenix in the Mazatzal Range	7659	Steep, rocky, little space on top. Profile on AZ license plates
Granite Mt.	NW of Prescott in the Bradshaw Mts., Prescott National Forest.	7295	Rocky, half covered with trees, fair incline.
Mt. Union	Highest mt. in the Bradshaws	7988	Road on W. side that may go to the top. Buildings at the top. Completely covered with low vegetation.

Weaver Mt.	Near Yarnell	6574	Barren. Rich in minerals.
Martin Mts.	In the Harquahala Mts	7073	Flat top, sparse vegetation, probably road to top, rich in minerals (i.e. mining).
Picacho Butte	Between Seligman and Ash Fork.	7168	About 1000 sq. ft. on top. Not enough room.
Mohon Mts.	40 mi. WNW from Prescott.	7499	Sloped top, barren, steep E side.
Mingus Mt.	10 mi. SW of Cottonwood.	7815	Heavily forested on top. Ranger station, microwave tower, road to top.

Kitt Peak was named by the local surveyor (Roskruge) in the 1870s for his married sister, Mrs. Philippa Kitt. The Tohono O'odham name for the mountain is Lolkam Du'ag, which means "mountain of manzanita shrubs".

I also looked at Cottonwood Peak (8889') just north of Mt. Graham and later I went up on horseback with a forest ranger from Safford. It had Ponderosa pine trees at the top. Beautiful site! With predominant winds from the northwest, the higher Mt. Graham would not affect it. In retrospect, one could develop Cottonwood Pk. (there is no road to the top) and perhaps develop Mt. Graham (10,720') later. But we never pursued that. Perhaps being farther east, the weather would be worse than at central Arizona sites. The Catalinas, Rincons and Santa Cruz Mts. were judged to be too close to Tucson lights, and Kitt Peak was a real possibility.

On the trip dinners were generally $1.70 and motels $7.14. Lunch at Nogales Airport from two vending machines was 5¢ for coffee for the pilot and 5¢ for a candy bar for me. Total 10¢. I was criticized for submitting such a low expense account because it would be unfair to others.

After three days I asked him how much I owed him. He estimated that we flew 2000 miles, so at 10 cents per mile for him, his plane and gasoline, I owed him $200. On the way back his engine started to sputter, so he landed on an uninhabited dirt strip in Texas E of El Paso and radioed his wife. She came from Marfa, but he found that it was due to a dirty sparkplug, which he cleaned. He flew back and I rode with Mrs. Casparis.

Later Aden and I added the Chevelon Butte (6945 ft., beautiful!) south of Winslow, the Sierra Ancha Mts. (Aztec Pk., 7694 ft.) southeast of Payson, Harquahala Mt. west of Phoenix, and a site north of Flagstaff. In October 9-11, 1955 Aden, Harold Thompson (draftsman, later Superintendent on Kitt Peak), and I drove from Yerkes to Phoenix in my 1955 red and cream Chevrolet convertible, bought in Marfa. A headquarters office was set up at 221 E. Camelback Rd. and Margaret Barker was hired as a secretary.

During the second year (1956) I went to various sites by Jeep and foot. For instance, I climbed the high peak in the Whetstone Mts. and camped overnight to see the city lights of Ft. Huachuca and the smelter smoke. After the above explorations with Aden, I returned to Yerkes to teach while testing was done of the most promising sites.

The testing consisted of three main parts (Meinel 1958). He designed two 16-inch Cassegrain telescopes with photoelectric photometers that fitted on trailers and were hauled to several sites (a site NE of Flagstaff, Chevelon Butte, Hualapai Mts., Sierra Ancha Mts., and Kitt Peak) where Claude Knuckles and John C. Golson did photometry of stars to measure the transparency of the sky. Also he designed 100-foot tall structures that allowed measuring temperature fluctuations at different heights above the ground. Then he designed triple 100-foot towers with automatic photometers at the tops that were pointed at the N pole and let Polaris (3/4 deg. from the pole) move around in a circle so that its light went through Rhonchi screens. If the seeing were perfect, the transmitted light produced a Roman-style pattern of full and zero light. If the seeing was poor, the pattern was a sine wave. Those towers were placed on the Hualapai Mts., Chevelon Butte, the Sierra Ancha Mts., and Kitt Peak. It quickly became obvious that the seeing on Chevelon Butte was poor because the winds from the northwest passed over a long flat plain and caused poor seeing as it passed over Chevelon Butte, only 600 ft. (200m) above the plain. The tower on Sierra Ancha Mts. had to be removed because it was accidentally placed on private land without permission. The final choice was between the Hualapai Mts. and Kitt Peak. Meinel's (1958) summary explains why Kitt Peak won on about 16 out of 18 criteria.

In 1958 Kitt Peak was selected by AURA to be the site of the national observatory. This had involved obtaining permission from the Papago Reservation Native Americans to grant use of the mountain top. The Papagos (later renamed the Tohono O'odham Nation) had been swindled numerous times by white men and their initial reactions to all proposals was to say "No." Prior to 1938, non-natives could dig mines on Native American reservations without permission but the *Indian Mineral Leasing Act of 1938* said that they had to obtain permission from the tribal councils and the Secretary of Indian Affairs. However, the Native Americans could not terminate or change leases or initiate leases themselves. Later the *Indian Mineral Development Act of 1982* gave the Native Americans full rights to the minerals on their land.

Fig. 12-1. The Kitt Peak National Observatory. The 2.1m Telescope (upper right) with the coude feed telescope projected just in front and left of the 2.1m itself. Most of the observations to show that most solar-type stars have companion stars were obtained with the coude feed.

In 1958 several Tohono elders were invited by Dr. Edwin Carpenter (Univ. Arizona) to see the Moon and other astronomical objects through the Univ. Arizona's 36-inch (0.9 m) reflector on the campus. They learned that an observatory on Kitt Peak would respect the environment and, like their own people, would passively explore the sky. Furthermore, Meinel and others promised to give Tohono O'odham (1) preferences to jobs at the observatory for which they were qualified, (2) sell (at no cost to the observatory) Native American handicraft, and (3) an annual stipend. Unemployment on the reservation was very high. With those stipulations written in, an agreement was signed between the Tohono O'odham and the NSF to last at least 100 years and as long as the facilities on Kitt Peak were used for scientific purposes.

In the summer of 1959 Aden offered me a position at Kitt Peak National Observatory. I accepted and moved to Tucson in August 1959. From then on I helped design, install, use, and instruct visiting astronomers in the use the telescopes and spectrographs. Aden and I made monthly trips to South Pasadena, CA to discuss details about the telescopes and spectrographs that Boller & Chivens was building for us. Each time when we found an error or correction, we felt that we had justified the cost of our flights and trips. I do not remember my positions at any one time or when I changed from being an Assistant Astronomer to Associate Astronomer (with tenure).

I enjoyed teaching visiting astronomers how to use the equipment, usually by spending their first half-nights with them. I remember often getting phone calls in my office during their observing runs with a problem, so I jumped into my car and was there in 1.25 hours to solve the problems. For that reason, plus my own observing, I kept an afternoon-evening work schedule (until about midnight) and do so to the present.

We installed a 36-inch (0.9 m) and an 84-inch (2.1 m) Cassegrain spectrographs of Aden's revolutionary designs (large collimator-to-camera focal lengths) and the 84-inch coude spectrograph with a provision for six cameras, of which three were built, all for photographic use. Later Judy Cohen converted one of the cameras to a solid-state detector and Art Hoag supervised the installation of a "coude feed" telescope. With a primary of 36 inches (0.9 m) and two reflections to the slit instead of five, the coude feed was nearly as fast as using the 84-inch and could be used while the 84-inch was used by others.

When Kitt Peak took over from Yerkes the project of an observatory in Chile that became Cerro Tololo Inter-American Observatory, I went there seven times to observe and install a 36-in (0.9 m) and a 60-inch (1.5 m) Cassegrain spectrographs and the horizontal coude spectrograph for the 60-inch (1.5 m) . Probably those are no longer used. Unknown to most people, the 60-inch (1.5m) telescope was paid for by the US Air Force while astronomer Gerhard Miczaika was working for it, but that fact was never publicized because it did not seem politically wise to say that the US military was helping fund CTIO.

It was logical to have the nascent observatory receive policy guidance from existing observatories by setting up an organization called AURA (Association of Universities for Research in Astronomy). (Incidentally, that name has been misspelled for several decades on the sign outside the front of the Kitt Peak headquarters building at 950 N. Cherry Ave. in Tucson, despite my pointing out the error several times.) AURA initially consisted of two representatives from each of the seven universities in the US that granted Ph.D.'s in astronomy in 1957. The representatives were generally the heads of the astronomy departments and a university administrator. Caltech declined to take part, but sent an observer (Dr. Ira Bowen). The current membership of AURA is 47 universities with one person from each. The increased membership meant that it could no longer meet in the KPNO conference room, where they could talk with staff astronomers; instead it met at expensive resorts.

The idea of organizing AURA to oversee KPNO seemed like a good idea but the organizers failed to foresee that a new major observatory would compete with the existing observatories for the funding (governmental and private) available for astronomy. That problem came to a head when Caltech and the Carnegie Institution of Washington applied to the Ford Foundation for funding for a large telescope in Chile to supplement in the southern skies the innovative research studies being done in the northern skies at the Mt. Wilson and Palomar Observatories and the University of California's Lick Observatory. The Ford Foundation was receptive to the idea but surprised everyone by awarding the money (for the 4m Blanco telescope) to AURA, rather than to Caltech-Carnegie-UC. The reasons why it did so was (1) NOAO already had an operating observatory there and (2) due to the ways in which the two consortia granted observing time.

In the California system each senior staff member was granted a fraction of the available observing time determined initially upon employment, regardless of the success or failure of his/her recent observing. In the AURA system each visiting and staff astronomer had to apply semi-yearly for telescope time by explaining in detail how the time would be used and the success of recent assignments. The occasional failure of the California system was epitomized by Halton Arp. "Chip" Arp had done excellent research on the color-magnitude diagrams of globular clusters, the variable stars in globular clusters, and the "peculiar" galaxies. But when he started to work on quasars, he failed to use scientific methods. That was during the early years when we did not know what quasars were. He searched in fields around nearby galaxies and found many quasars surrounding them. He assumed - without evidence - that the galaxies and quasars were at the same distance from us, even though their redshifts were very different. He thought that the quasars were shot out of the nearby galaxies and concluded (1) that the redshifts were not Doppler shifts and (2) that major changes were needed in physics to explain the difference. But what the referees of his papers and the Editor of the ApJ told him is that he needed to do is a control experiment, namely to look at a field that has no nearby galaxy in it, and find out if it has the same number of quasars per square degree. He replied that he was not given enough observing time to do a control. So eventually, but not soon enough, he was deprived of all observing time. He resigned his position at the Carnegie Observatories and moved to Europe, where his wife had a position at ESO. So sad – because he neglected basic principles of science.

The members of AURA were influenced by the California observatories and held back KPNO and CTIO. Meinel had plans and optical deigns for a telescope larger than 4m, but was not allowed to proceed by AURA dictate. Roger Lynds was working with people (e.g. Dick Aitkens) who were producing much more sensitive CCDs than other companies and Roger was able to get spectra of objects at 21st magnitude. He resolved the Lyman-alpha forest that showed the existence of galaxies being formed early in the lifetime of the universe along the light paths to quasars. He also obtained spectra of the relativistic rings around galaxies due to the gravitational bending of light because of quasars along the light paths. He was thus able to prove that, at times, the national observatories could do better observational research than the California observatories. That only enhanced the competition and bad feelings.

KPNO was initially able to talk directly to the NSF (National Science Foundation) regarding needs, but then AURA, with its mixed loyalties, intervened. With the establishment of CTIO (and later other observatories) AURA felt it necessary to insert another administrative entity between the observatories and AURA. Those organizations, AURA and NOAO, wanted to reap credit for the discoveries coming out of CTIO and KPNO, so the observatories were never allowed to have press officers or distribute press releases. AURA and NOAO wanted all the credit for the discoveries, even though they consisted mostly of administrators. When one looks at the KPNO headquarters building in Tucson nowhere on the outside does one see "Kitt Peak National Observatory", only AURA and NOAO. On the tower above the optical shop are three signs saying NOAO, LSST, and DKIST (a telescope in Hawaii) but not KPNO.

The third demotion of KPNO came when NOAO wished to include the Gemini Observatories (in Hawaii and Chile), the National Solar Observatory, and LSST (Large Synoptic Survey Telescope). Then NOAO was abolished in favor of a new organization called "National Optical and Infra-red Research Laboratory." The acronym for that is NOIRLab, which in French means "Black Lab". That is the antithesis of observatories that collect light. A simpler name would have been "National Observatories", but we were not asked for suggestions. This required another administrative organization that siphoned off an additional part of the funds for astronomy. Now with the McMath-Pierce Solar Telescope and coude feed telescope closed because of a lack of funds, and the Solar Vacuum Telescope destroyed, and with most of the observing time on the 2.1m, WYIN, and 4m assigned to three large projects, there is virtually no observing time left for individual American astronomers. Unless an astronomer's research interests is narrow enough to be satisfied with one of the three remaining instruments on Kitt Peak, there is little observing time left on Kitt Peak for American astronomers. The NSF went through a decade of lean budgets and had to reduce the KPNO budget. It reasoned that there is enough observing time for individual astronomers on the Gemini and CTIO telescopes, the Steward and McDonald Observatories, the Sloan Digital Sky Survey, the space telescopes, etc. that KPNO is no longer needed. Therefore, the original goal of the 1950s of a national observatory on Kitt Peak that provided a wide range of instrumentation for all American astronomers to use has been effectively abolished.

10 LIBRARIES

Before WWII there were two locations where research papers were published: in observatory publications and astronomical journals. Roughly half of the published papers went into each, but they were supported financially in different ways. For the observatory publications the total cost of publication and distribution was paid by the observatory. The publications were sent free to the other observatories, in exchange for their publications, and given free upon request to individual astronomers. For the journals the costs of publication of the initial copies were paid by the publishers and recipients paid for the costs of the additional copies and their distribution. Eventually the observatory publications disappeared in the 1970s and 1980s because they were too much of a financial burden on the observatories. Also the observatory publications lacked outside reviewing, so in controversial areas they lacked objectivity.

I was placed in charge of building libraries for four locations: Tucson, Kitt Peak, La Serena, and Cerro Tololo. While we could buy some series of journals from used book dealers, they seldom offered sets of observatory publications. I discovered that the best and cheapest source for the observatory publications and journals was in the private libraries of astronomers. Each senior astronomer had a private library for his own use, having gotten the observatory publications free for the asking. People joked that the first person to contact a recent widow after the undertaker was me. I made generous offers for the private libraries and saved the widows the chore of knowing what to do with those books. Thus we bought the libraries of Otto Struve, Alfred Joy, Gustav Strömberg, and many other astronomers. Duplication was not a problem because we could use up to four copies of each volume. Thus the KPNO library became one of the best in the country.

I also compiled a large personal library because as Editor, I needed to refer to referenced publications a lot. It saved thousands of trips to the KPNO library. But in the digital age printed books lost their value. Fortunately in 2000 after giving up the editorship, I learned that Tsinghua University in Beijing was building up its research in astronomy and its library. I donated 1200 volumes, roughly half bound, to Tsinghua University and it paid for shipment to Beijing. On one trip to China I saw the collection in their astronomy department; it was totally bound.

However as a warning, not all of the older papers are available online. In a paper that I published in 2018, I counted that 37% of its many references were not available online. As Einstein said "The only thing that you absolutely have to know, is the location of the library."

11

GORDON GRANT

Gordon Grant was an undergraduate at Case Institute of Technology in the 1950s and a NSF Predoctoral Fellow in 1955-6. He was energetic, light-hearted, humorous, and a joy to be around, but not a near-genius as is required for success in astrophysics. He published papers with Victor Blanco and Lawrence Aller. He did graduate work at the Yerkes Observatory of the University of Chicago and got a Ph.D. under Gerard P. Kuiper in 1958 for studies of lambda Tauri and RW Tauri. They were based on photoelectric and spectroscopic observations he obtained at the McDonald Observatory. Then he took a position in Cleveland for a commercial company and moved there (at 2873 Ludlow Rd.) with his wife Patricia. Tragically, one day he went into the woods and hung himself. We can guess that he was losing his job and no longer getting along with Pat because he left no note. I finished four of his papers and saw them through publication.

12

SPECTRAL ATLASES AND CLASSIFICATIONS

12.1 ATLASES

Morgan and Keenan started the two-dimensional (MK) classification (temperatures and luminosities) system of stars and their Morgan et al. (1943) atlas was the standard source for classification. That atlas used prismatic spectra, whereas current spectrographs were all using gratings, which have nearly a constant dispersion with wavelength. Also it was realized that Kodak emulsions had resolutions of about 10 microns, rather than the 20 microns previously supposed. Meinel designed grating spectrographs with better resolutions for Kitt Peak, so it seemed appropriate to publish a new atlas.

A temporary atlas by Abt et al. (1968) was first published and a final excellent one by Morgan et al. (1978) was published later. I have all of the remaining copies of the latter, available free to anyone. Many of those original spectra were of 39 Å mm^{-1} dispersion and taken with the 84-inch (2.1 m) Cassegrain spectrograph. Those sheets compared stars with different spectra, temperatures and luminosities, both for normal and abnormal types. It is still the standard for photographic spectra.

12.2 LARGE COLLECTIONS

I continued to classify many stars. For instance, I started to classify thousands of members of visual multiple systems because sometimes unusual stars are in physical systems with normal stars and that allows us to learn the luminosities and masses of the unusual stars. For instance, while getting spectra of beta Lyrae, I noticed that there was a small group of stars around it. By the 1960s there had been more papers published about beta Lyrae than any star in the sky except for the Sun. So I convinced Hamilton Jeffers at Lick to obtain recent positional measures, Allan Sandage to obtain photometry, and I obtained spectral classifications of the members of that small group. We found that it was a physical group and obtained the absolute magnitude and mass of beta Lyrae, values that are still the best to date (Abt et al. 1962).

An interesting side story concerns Hamilton Jeffers, who was working at Lick Observatory when I was there. He had decided to produce a catalog of all the published double-star observations. He brought Willem Hendrik van den Bos up from South Africa, who was an expert on the southern stars. With the help of Rete Greeby, they produced the catalog on IBM cards, which was published as the *Index Catalog of Visual Double Stars*. That catalog was subsequently maintained at the US Naval Observatory as the *Washington Double Star Catalog*, mostly by Charles Worley. Hamilton Jeffers continued to make observations with the 36-inch refractor, but usually spent week-ends with his brother, poet Robinson Jeffers, in Camel-by-the-Sea. Brother Robinson popularized the Big Sur region of California and was an early environmentalist. Recently while I was in California I visited Robinson Jeffers' Tor House in Carmel-by-the-Sea which he built by hand on the Pacific coast. When I mentioned to the docents that I knew Robinson's brother, the cameras came out and a long interview occurred.

In large collections of spectral classifications (Abt 1981a, 1985b, 2008), I classified 2411 members of visual groups selected from the Aitkin (1932) double-star catalog. I selected all members of visual multiples brighter than B = 8.0 mag. In the first paper, I found that 1% are weak-lined stars, 32% of the A stars are Am, and 55% of the A3-A9 stars are of luminosity classes III or IV. In the second paper I found 12 Ap, eight λ Bootis, one Ba II, one SB2, and 60 Am stars. In the third paper I discovered 15 Ap stars, 33 Am stars, 18 composite spectra, three shell spectra, and two SB2s.

My last large collection of spectral types was of spectroscopic binaries. Allan Batten had published several catalogs of the orbital elements of binaries but the ninth one was by Pourbaix et al. (2004). It is available online only at http://sb9.astro.ulb.ac.be, which is continually updated. It lacks certain data. For instance, one-third of the AF stars lacked MK (Morgan-Keenan or two-dimensional) types. I published (Abt 2009a) MK types for 145 systems, whose data are available in the online publication.

12.3 STARS IN CLUSTERS AND ASSOCIATION

Although there were many color-magnitude (Hertzsprung-Russell) diagrams of open cluster and associations, spectral types of individual stars in the MK system were seldom known. A large collection of one-dimensional types of stars in clusters was obtained by Robert J. Trumpler during the last 30 years of his career, but that collection was never published by his son-in-law Harold Weaver. I had offered many times to finish and publish that collection, even without my name of the title page, but Weaver always had plans to do so himself.

With the ability to obtain the types of the brighter cluster stars and with the help of students and Dr. Hugo Levato of Argentina (who spent several years working with me in Tucson), we published spectral types for 12 clusters (alpha Persei, Coma, Orion Nebula, Pleiades (shown in the logo on Subaru cars), Ursa Major stream (including the Big Dipper), IC 2602, IC 4665, M34, M39, NGC 2169, NGC 6475, NGC 6633) and two associations (Lacerta OB1, Orion OB1). One goal of those studies was to discover abnormal stars in clusters of different ages. A summary of the results was published in Abt (1979), as explained in the following paragraphs.

The Ap(Si) stars do not occur in clusters younger than 10^6 yr, but by 10^8 yr they have the same frequency as field stars. The Ap(Hg,Mn) stars do occur in the first 10^7 yr, but by 10^8 yr they have the same frequency as field stars. The Ap(Sr,Cr) do not occur in 10^8 yr but by $10^{8.3}$ yr they have the same frequency as field stars. The latter explains why there are normal slow rotators in the field that have not yet become Ap stars. The Ap(Si) stars have decreasing rotational velocities with age probably because of magnetic breaking. The Be, shell, and sn stars show no variations with time; they have the same frequencies as field stars at all ages.

Morgan and Lodén (1966) called attention to the observation that many stars in the Orion Nebula cluster have unusually broad hydrogen lines. In the spectral atlas by Morgan et al. (1978), those with unusually broad hydrogen lines are called luminosity class Vb, while stars with normal hydrogen lines are called Va. We found that 50% of the brighter stars (-3.1 mag. $\leq M_V \leq$ +0.8 mag.) in the Orion Nebula cluster (age of $10^{5.7}$yr) are Vb, while such stars are very rare among older clusters.

The Am stars occur in the youngest clusters, telling us that it takes less than $10^{5.5}$ yr to produce them. The rotational velocities of Am stars decrease with age at the rate of $T^{-1/4}$, probably due to tidal braking in close binaries because essentially all Am stars are in short-period binaries. The rotational velocities of Am stars in the youngest clusters is ~250 km/s, which is five times that in older clusters.

I have read few observational papers that contained that many significant new observational results. I did all the work in developing the results in the Abt (1979) paper, but later regretted that I did not add Levato as a coauthor.

The above results depended on our having derived the rotational velocities of stars in at least five clusters and the binary frequencies in more than eight clusters. Those results were summarized in an *Annual Reviews* paper called *Normal and Abnormal Binary Frequencies* (Abt 1983b). For instance, I have been one of the few astronomers with enough courage to study the binary frequency in broad-lined stars. Garmany et al. (1980) found a deficiency of O stars with low-mass companions. Abt & Levy (1978) found that while long-period (>100 yr) binaries have a frequency of secondary masses like in a van Rhijn or Salpeter distribution, those with shorter periods have decreasing numbers of companions with decreasing masses.

13

VISITING PROFESSOR PROGRAM (1960-1962)

In the late 1950s there was only about two dozen North American universities that had departments of astronomy or physics & astronomy. The American Astronomical Society (AAS) was concerned about attracting science and technology students to careers in astronomy at the other colleges and universities. Therefore in 1959 it set up a Visiting Professor Program in which any 2-year or 4-year college could have an astronomer visit for two days to give talks on contemporary astrophysics and what is involved in becoming as astronomer. The host college paid a modest $300 (if it could) and the AAS paid the travel & living expenses and stipends to qualified astronomers. The program was supervised by AAS Executive Officer H(erman). M. Gurin. The first year the only astronomers were Paul Merrill, Seth Nicholson, and Harlow Shapley. Then it was thought to be more productive to send a younger astronomer, so I was selected to be the sole Visiting Professor in 1960-2. Later the program was renamed the Harlow Shapley Lectureships and more astronomers became involved.

My calendars list the 15 colleges (there may have been more) that I visited, which are listed in Table 2.

TABLE 2: VISITING PROFESSOR PROGRAM VISITS

Dates	College
1960	Baldwin Wallace Univ., Cleveland, OH
	Jamestown College, Jamestown, NY
	Kent State Univ., Kent, OH
	Jackson State College, Jackson, MS
1961 Feb. 9-10	Grand Canyon College
Mar. 13-14	San Diego State Univ
Apr. 24-25	Lewis & Clark College, Portland OR
Apr. 26-27	Reed College, Portland OR
1962 Feb. 9-10	Univ. New Mexico, Albuquerque, NM
Feb. 13-14	Midland College, Midland, TX
Feb. 15-16	New Mexico State Univ, Las Cruces, NM
Mar. 15-16	Black Hills State Univ., Spearfish, SD
Mar. 19-20	Washington State Univ., Pullman, WA
Mar. 21-22	Idaho State Univ., Pocatello, ID
1963	Texas Women's University, Denton, TX

My visit to Kent State College was 10 years before the killing (on 4 May 1970) of four students who were protesting our participation in the Viet Nam war. And to Jackson State College 10 years before two students were shot on 15 May 1970.

In 1960 Jackson State College was a black-only college meant to be for blacks what the University of Mississippi was for white students. The administrators at Jackson State freely admitted that the blacks coming from black high schools were inadequately educated for college, but (1) by the time they graduated from Jackson State they had the equivalent of a good high school education and (2) they qualified for federal positions that required college degrees. I freely accepted invitations to eat in their cafeteria, but the teachers were nervous about the possibility of a white tradesman seeing a white person eat in the black cafeteria.

Each college was different and I hope that I turned some students to think about careers in astronomy, but never learned whether it did. It was in these talks that I became much at ease in talking to audiences (up to 400), sometimes with no advance notice. I gave up to nine talks in two days. Later the AAS hired many astronomers to do this, each one visiting only a few colleges near to them.

14

FREQUENCIES OF DOUBLE STARS

14.1 METALLIC-LINE STARS

The metallic-line stars (abbreviated Am) are main-sequence stars that have weak Ca II H&K lines, indicating types of A2-A6. They have hydrogen line strengths of A7-F2 that are typical of stars with the same colors and temperatures. They have strong metallic lines as in A8-F5 stars. Greenstein (1948) found that their atmospheres have low pressures and gravities and high turbulence like in supergiants. He also found that the abnormal abundances are among the metals whose second ionizations are around 13 ev, the ionization energy of hydrogen. Slettebak (1954) found that all Am stars have low rotational velocities, namely V sin i < 100 km s^{-1}.

To this collection of strange characteristics, I added one more. I studied 25 Am stars for duplicity and found that at least 88% are double (Abt 1961). Because that was done with radial velocities, a few doubles are missed because their orbits are face-on. So apparently, all Am stars have companion stars. How can one explain this set of odd behaviors?

A clue to these behaviors came from a study of the normal A-type stars that occupy the same part of the main sequence as the Am stars. In a study (Abt 1965) of 55 normal stars, I found that 17 out of 55 stars were double stars, but all had periods greater than 100 days. After corrections for the aspect effect, I found that Am stars have periods of 0-100 days while the normal A stars have periods of 50-250 days. Thus if an A star is a slow rotator, it becomes an Am star and if it is a fast rotator, it becomes a normal star.

To understand this difference between Am and A stars, we have to consider the interiors of stars. There is a huge amount of material in a star, not just billions and trillions of tons, but numbers most people do not know, such as quadrillions, quintillions, and much larger. When all that material rests (due to gravity) on the core of a star, the core becomes very hot due to that pressure. The core temperatures are typically about 10 million degrees. Temperatures are not measured with a thermometer, which would evaporate, but by how fast particles are moving. At such high temperatures, the particles – mostly protons – are moving so fast that they take part in nuclear reactions, such as four protons combining to form an alpha particle. But an alpha particle weighs slightly less than the four protons, so the lost material turns into energy in the famous equation $E = mc^2$. That energy seeps out through the star and when it reaches the surface, it produces the light that we see.

But how does the energy move from the core to the surface? There are two ways. One method is called convection in which hot currents move outward and cooler ones move downward, so there is a lot of circulation. The other method is by radiation in which hot atoms farther down radiate light to cooler atoms farther out, and there is little motion of the material. In A stars there is a radiation zone just below the surface. With little circulation of the material, some atoms fall inward because they are heavier and other atoms float outward. It is called diffusion. The small motion causes a slowly-rotating star to make the hydrogen atoms fall inward and the metals to move outward into the surface regions, producing typical Am star spectra. But if the star is rotating rapidly, there is no radiative zone below the atmosphere and the diffusion does not occur, causing the star to have normal abundances of the atoms in the surface layers. This diffusion process was explained by Michaud (1970) and it explains why Am and Ap stars occur among slow rotators while rapidly-rotating stars have normal abundances.

If a star is in a binary or double-star system, the companions that are close cause the stars to rotate more slowly by tidal interaction. So that is why the Am stars are slow rotators and hence have strange abundances by diffusion while stars with no close companions, or only with distant companions with orbital periods greater than 100 days, continue to rotate rapidly, have diffusion that eliminates the radiation zone, and have normal abundances. Complicated? But this makes good physical sense.

There is one remaining problem that I solved in Abt (2009b). I said that Am stars have rotational velocities of 0-100 km/sec and normal A stars have rotational velocities of 50-250 km/sec. That means that stars with $V = 50$-100 km/sec can be either Am or A. However, studies of stars in open clusters show that it takes several million years for the diffusion mechanism to take effect. Therefore once a star reaches the main sequence, it will be a normal A stars for several million years before it becomes an Am star.

14.2 THE Ap STARS

Ap (or A peculiar) stars are ones with extremely strong lines of unusual elements. They fall into three sub-groups. The Ap(Hg,Mn) stars have strong lines of mercury (Hg), virtually never seen in any other stars. The Ap(Si) have strong lines of silicon and the Ap(Sr,Cr,Eu) stars have strong lines of those elements and of the rare earths. Babcock (1958) found strong magnetic fields of ~5000 gauss (compared with the earth's magnetic field of 0.2 gauss) in the latter two groups but not among the Ap(Hg, Mn) stars.

Abt & Snowden (1973) studied 62 Ap stars and found a normal fraction of binaries (40%) among the Ap(Hg, Mn) stars but only 20% among the other two groups.

15

JAPAN

15.1 JAPAN IN 1965

Kitt Peak had several visiting astronomers come from Japan and from what I had read about Japan, I decided to go there. I invited Pat Osmer, who had worked for me during three summers, to go along. He was a student at Case Institute of Technology. Then he did graduate work at Caltech (I helped him to stay with the Bolenbaugh's and work for Allan Sandage) and obtained a Ph.D. He got a position at CTIO and eventually became its Director. Finally he went to Ohio State University, became head of the Astronomy Department and University Provost.

We went to Japan during August 2-24, 1965. The plane stopped in Honolulu. I thought that Hawaii was just a tourist trap, but decided to stay one day. Boy, was I wrong! I fell in love with Hawaii in about 5 min. Simply getting off the plane and smelling the strong smell of flowers did it. We stayed in a hotel near Diamond Head that was so informal that we could eat in the restaurant barefoot and in our swimming suits. Swimming in the ocean was warm and pleasant. In later years I returned to Hawaii nearly a dozen times, usually staying with the Ann Boesgaard (an astronomer at the Univ. Hawaii during her entire career) and her husband Hans (Chief Engineer at the Univ. of Hawaii) and sailing on their sailboat to neighboring islands.

Kiyoteru Osawa met us at the airport in Tokyo and made reservations for us to stay at the visitors' center of the Tokyo Univ. We had dinner with him, who is an astronomer at the Tokyo Observatory, and with Jun Jugaku, Editor of the Japanese Journal of Astronomy. Jun had previously worked at Caltech on Jesse Greenstein's stellar abundance NSF grant. Dinner was good but the locust outside sure were noisy! The next two days we toured Tokyo and had dinners with the two. After the trip I sent Osawa a copy of *The Bright Star Catalogue*.

I rented a car – a Toyota Corolla. Pat and I wanted to climb Mt. Fuji (12389 ft. = 3776 m), a semi-active volcano (the last eruption was in 1707). Jun was not convinced that we could find the way (100 km from Tokyo), so he drove there himself and we followed. After arriving, he drove back to Tokyo. After the trip I sent him a book on stellar structure.

The "thing to do" was to hike up the mountain during the moonlit night and see dawn from the top. We were not the only ones wanting to do that: there were about 40 buses at the bottom and on the trail up people were 4-5 abreast! The view from the top was spectacular; it is high above everything else in Japan. Going down we slid down a cinder shoot. That was illegal (we didn't know that at the time) but fast. I lost my glasses; I had been wearing glasses for nearsidedness since the age of 8, but realized that I did not need glasses for distance seeing.

Then we drove to Kyoto, stopping for a swim at Atsumi. There was a small island offshore, so we swam to it. Most Japanese could not swim, so they were amazed to see us do it. They also do not like to get tanned, so why do they go to the beaches? To eat!

In Kyoto we stayed at the Daisen-in ryokan (a Japanese inn with bamboo mat floors, bedding stashed away until nighttime). From the small balcony a stream flowed past. It had a common unisex showerroom. We visited the Silver Pavilion, Golden Pavilion (see Fig. 15-1), and Nara.

Then we drove to Okayama Observatory, which has a 74 inch = 1.88 m telescope, and met (again) Goro Ishida, Director. Assistant Paul Misawa drove us to a ferry to Shodoshima Island, the second largest in the Inland Sea south of Honshu. Then another ferry to Shikoku Island, which has the amazing Naruto Whirlpools at the east end. The currents around the island meet at the east end to cause a continuous whirlpool that looks dangerous.

Fig. 15-1. The Golden Pavilion (Kinkaku-ji), a Zen temple in Kyoto. The upper stories are covered with gold leaf. It was burned down in 1950 by an acolyte who confused his love for the temple with his stuttering, as told by Yukio Mishima in his book The Temple of the Golden Pavilion. The temple has been rebuilt.

From there we drove to Fukuoka, the main city at the north end of Kyushu Island. We met Dave Woods, whom we knew in the States; he was at the American Itazuke Air Base there. He took us partly up Mt. Aso (5223 ft. = 1592 m), an active volcano. The last eruption was on 29 July 2019 and people are currently not allowed on the mountain. We did not see Nagasaki.

The next day we drove to Hiroshima and Osaka, where we took the "bullet train" to Tokyo. It was the fastest modern sleek train of the time. We did some shopping there. I am especially glad that I bought a signed woodblock print by Kiyoshi Saito of a snow screen in Hokkaido, the northern island.

15.2 JAPAN IN 1990

Kam-Ching Leung (Univ. Nebraska) obtained an NSF grant for a joint US-South Korea astronomy meeting. It included funding for travel by 12 American astronomers, namely Leung, Abt, Anne Cowley, Ed Guinan, Robert Honeycutt, Robert Koch, Yoji Kondo, Hal McAlister, Ron Taam, Ron Webbink, Bob Wilson, although Icko Iben dropped out in favor of Dan Popper. Arcadio Poveda from Mexico met us. This became then the first in a series of conferences called the Pacific Rim Conferences on Stellar Astrophysics. Back in Tucson, Jay Gallagher promised to come there for 3-4 days to handle the ApJ while I was gone (Nov. 3-17, 1990). I arranged with Goro Ishida to visit Japan for a couple days on the way back.

I flew Northwest flight 23 to Seoul. Seoul was the 4th largest city in the world at that time after Shanghai, Mexico City, and Tokyo. South Korea had 40 million people, of which 10m were Buddhist and 10m Protestants.

Skipping ahead to the Japanese part, Jun Jugaku met me at Tokyo airport on Tuesday, Nov. 14, two hours' drive from Tokyo. He couldn't drive because of the Monday coronation of Emperor Akahito and some roads were closed the following days. We stayed the airport Holiday Inn. I gave him a Sierra Club calendar and a smoky topaz necklace for his wife. The next morning we took the hotel bus to the airport. There were many police with shields because of the farmers' protests for the government taking their land for the new airport.

The next day was perhaps my most interesting day abroad that I have ever had. First, Goro Ishida and I went to the national museum to see a collection of kimonos (colorful, beautiful) of all periods. It overlooked the Imperial Palace garden. Then we walked along the Ginza to the Kabumiza Theatre that shows Kabuki plays. We saw one hour-long act of the 8-hour play, the one that has the famous Lion Dance. In that dance, a reincarnated women has a long white mane that she swung around 36 times to the cheers of the audience. Kabuki was started in the 17th century and is very popular today. It involves singing and dancing. . The singing is done by two rows of actors with bamboos flutes and samisens in the back and drums and one flute in the front.

Then Goro took me to the Kabuki dressing room and we had tea with Izaemon XVII, one of Japan's 85 "National Treasures". National Treasures are individuals or groups who are outstanding in cultural, artistic, scientific, educational, or technological areas, such as Gagaku, Noh, Kabuki, music, dance, drama, ceramics, crafts, etc. Izaemon XVII's name is a traditional stage name, like those of Popes. He (~65 years) played female roles; face masks are not used in Kabuki. Goro knew him personally because he once wrote to him asking why he made such-and-such a motion when he said…. in one role. So they continued their correspondence. Goro knew Kabuki very thoroughly!

Next we went along the Ginza with its extremely colorful lights, crowds, and stores. We went to a restaurant known for its baked eel - good salty spiced mild fish on rice. Then we went on three different trains to the National Noh Theatre, built about five years previously with a modern exterior and a traditional style inside. The audience was of about 300, of which I was the only westerner. The Noh plays were started in the 14th century by Kan'ami and his son Zeami. The plays involve actors reciting stories, but not acting them out. It is very stylized. There are roughly five musicians. I have read many Noh plays because the Univ. Michigan published translations of many of them.

The first Noh play was called "Oba Ga Sake" about a nephew who tricked his stingy aunt into giving him sake (rice wine) for the first time in her sake shop. The second Noh play was taken from "The Tale of Genji" (which is the world's oldest novel, 1100 pages long and written in 1000 AD by Lady Murasaki Shikibu). The play is called "Nonomiya". It was played by a maiden (wearing a mask), priest with an excellent voice, villager, and a chorus of eight. Also three musicians playing a bamboo flute and two drums. One drum was hit by hand and one with hard pieces on the fingers. One drummer sang wordless sounds. Plus, three attendants. The maiden performed two stately dances. Then

Goro and I took a train to my hotel: Keio Plaza Hotel, where Goro, Jun Jugaku, and Tanaka said goodbye to me.

15.3 JAPAN IN 1997

In 1997 I decided to go to the IAU General Assembly in Kyoto, Japan. Alan Batten asked me to talk on *Access to Journals* in a joint discussion on developing countries. Then there was a meeting on "Stellar Astrophysics" in Hong Kong to call attention to astrophysics research efforts by the various universities there. Sidney Wolff agreed to pay my expenses. I paid for John's airfares with me. Then Kam-Ching Leung and I were invited to become Guest Professors of Peking Univ. by Qiao Gudjun, so he asked us to come to Bejing after the Kyoto meeting. He paid for our expenses in China. Hong Kong paid for my registration and housing in Hong Kong.

John prepared two talks for China. I prepared the following talks:

1. *Disks that Appear and Disappear around Rapidly-Rotating A Stars* (H.K., China)
2. *Access to Journals* (Kyoto)
3. Some notes for the Beijing ceremony
4. *The Formation of Binaries* (China)
5. *The Lifetimes of Trapezium Systems* (China).

Anne Cowley and Steve Shore agreed to substitute for me on the ApJ. Aug. 10 started out sadly because it was obvious that Kitty Kat, my cat for 21 years and 3 month, had to be put asleep because he was unable to get up off the kitchen floor and lay in his urine. We took him to Pima Pet Clinic. Both Daniel and I cried.

Skipping ahead to the Japanese part, I had made reservations to stay at Matsubaya Ryokan in Kyoto, starting 18 August. It was a pleasant room (No. 10) with a bamboo mat floor; the servants laid out our bedding on the floor in the evenings. Outside a wide open door was a small pond with carp swimming in it. We loved to sit on the edge of our room, eat our dinners, and dangle our legs toward the pond. In the communal bathrooms the Japanese retract their foreskins, a practice called Mie Muki, because they think that it is "uncool" not to do so. John and I treated them to the American practice of having them removed entirely, i.e. circumcision.

Not knowing any Japanese, ordering a meal in a restaurant could be difficult. However, many restaurants have a glass case in front with models of the various meals and drinks. One motions to a waiter to come out and point to the things one wants.

John and I visited several temples. One was SanjuSangendo Temple, built in 1164. It has 1000 gilded statues of Buddhas. But my feeling was that quantity does not equal beauty. We went to one of our favorites, the Zen Buddhist temple Ryoan-ji with the large sand garden. One can sit for a long time on the edge of the temple and enjoy the abstract beauty of the garden. Last, we went to Shokoku-ji Temple, built in 1383-1392 but burned several times. It was on the grounds of Kinkaku-ji (Fig. 15-1), commonly called the Golden Temple. After those, John visited many more temples in Kyoto (e.g. Nishi-Hongan-ji, To-ji, Tofuku-ji) while I went (20 Aug.) to the International Astronomical Union (IAU) General Assembly, held every three years.

The IAU General Assembly normally brings in >3000 astronomers from all over the world. I met friends from Mexico, Denmark, Japan, US, South Korea, Argentina, etc. The welcoming address was given by Emperor Akihito and Empress Michiko.

On 21 Aug. John and I visited the Kyoto Imperial Palace, having made prior reservations. There were gates for royalty, sub-royalty (e.g. Prince Charles), etc. The Ceremonial Room has a throne and 18 steps up to it. The colors were mandarin orange at the left and cherry on the right. In the royal quarters, the Emperor's bedroom was of the left, the Empress' on the right. Outside in the garden there were three teahouses.

On 24 Aug. we took the subway and train to a "Kei Seki" restaurant to meet Jun and Kuzako Jugaku, Jean-Claude Pecker, Brian Warner, and Kazuko & Saburo Ishida. Goro Ishida had died a couple years earlier of a heart attack. The 10 exotic courses consisted of mushroom, eel in seaweed, fish eggs & rice, etc. Kozuko Ishida gave me a bowl marking Goro Ishida's asteroid 5829 and a woodblock painting by Yuji Watabi (born 1974). It shows several people holding a large cloth to catch a comet; it is now in my guest bedroom.

On 25 Aug. both John and I went to the IAU General Assembly. He went to the symposium on the History of Oriental Astronomy and I went to the one on Electronic Publishing. The next day I went to the General Assembly symposium on developing countries. John had visited four more temples, including Nanzen-ji, Chion-ji, and Yasaka Shrine. We received two more FAXs about Mother having fallen a second time; Karl and Donna had flown to Tucson and were staying at my house. The following day John and I started back, taking a bus to Kansai Airport, built on reclaimed land. I phoned Mother and then Karl, learned that Mother had fallen a third time and was in pain. They were taking her to her doctor to find out if she had any permanent damage. Karl and Donna would stay until Sep., when we would be back from China.

16

THE UNUSUAL SCENERY IN ARIZONA AND UTAH

The US has beautiful and strange scenery, particularly in the West. Allan Sandage and I loved the Muir Trail (211 miles = 340 km) and hiked most of it in sections, namely all the way from Kings Canyon National Park to Yosemite, in the 1950s. Allan tried his hand at fishing while I explored the lakes, such as in the Hundred Lakes Basin. But what really turned me on were the parks and fascinating places in Arizona and Utah.

It started with my first visit and hike down into Grand Canyon in 1949, described in Section 5. Compare that with the current crowds that require reservations weeks in advance to hike in the canyon.

I hiked in the Canyon a dozen times since. I hiked across the canyon twice, which takes three days. Also I hiked to Clear Creek twice. That is the next stream flowing into the canyon from the north side and east of Bright Angel Creek. It involves hiking across the Tonto Plateau for ~15 miles, avoiding rattlesnakes. It had a lovely stream in a pleasant valley with cottonwood trees. However the second time (with Bill Reid) was after a flash flood that had wiped out everything in the valley, leaving a bare 20-foot wide by 10-foot deep rectangular rocky channel.

A lovely place in the Grand Canyon is Havasu Canyon, about 190 mi. west of Grand Canyon Village. It is the site of a small tribe of Native Americans called the Havasupai. There is a road that leaves highway US 66 at Peach Springs to the trail head (63 mi. away), called Hualapai Hilltop. Hiking into the canyon is an easy 8-mile hike to the village of Supai and two spectacular waterfalls. I went there several times with Peter Moseley, Andy Bacher (son of Robert Bacher) and others. Once Fred Chaffee and I decided to hike all the way thru the canyon to the Colorado River. I was Fred's Ph.D. thesis advisor at the University of Arizona. He became Director of the Smithsonian Observatory on Mt. Hopkins and did such a good job that he became Director of the Keck Observatory on Mauna Kea for many years. The hike involved crossing the stream many times, but we were prepared for that.

Later trips into southern Utah were mostly in my Jeep. We (Peter Moseley, Rich Bolenbaugh, Pat Osmer, and others) entered Canyonlands (decades before it was a National Park). The Jeep trail through the Park and as far as the Fins, beautiful Chesler Park and out to the south (to highway 95) would challenge any Jeep driver. Now Jeeps are not allowed into the Park; only hiking is allowed (Fig. 16-1).

A fascinating region on the west side of the Colorado River, across from Canyonlands, is called The Land of Standing Rocks and is now included in the Glen Canyon National Recreation Area. The way in was to take the ferry across the Colorado River at Hite, than a bridge across the Dirty Devil River that was built in the 1950s, and then northeast roughly parallel to the Colorado River. From the Land of Standing Rocks there is an easy hike down to the Colorado River and up the east side. That route was used in the 19th century by cattle rustlers who took stolen cattle away from the ranches around Hanksville. The Land of Standing Rocks has tall (100 ft.) monuments and small canyons, as Peter Moseley, I, and others found.

Fig. 16-1. "Newspaper Rock" in Indian Canyon on the way into Canyonlands National Park. Most of these petroglyphs date to ~1000 years ago; some are more recent. They were scratched into the natural desert varnish.

My favorite place to go hiking was in Zion National Park, especially the West Rim Trail. Many times I drove to Kanab, checked into a motel with a pool, had a good dinner, and went to the play in which some college student performed every summer evening. Later they moved their summer theatre to Cedar City. I got up early the next day to be on the trail by 6 am so that I was at higher elevations before it got hot. The trail had a spur to Angel's Landing, but I was too frightened of falling to do that part. Then the trail went over the broad ridge (Horse Pasture Plateau), passing sparse woods and clearings. Finally it comes to a mysterious canyon (Wildcat Canyon?) parallel to the main canyon of Zion. Wildcat Canyon has strange monuments and formations. I never hiked down into it; I simply sat on the edge looking down into it. Once I hiked some miles north along the ridge to the edge of a grassy meadow between the forests and camped for the night. I cooked my dinner quietly while deer grazed in the meadow. I never met a person on the West Rim Trail.

On one trip to Zion I passed the Olympic torch bearer on the lonely road between Page and Kanab (57 miles) at the foot of the Vermillion Cliffs. There was a travel trailer every 10 miles or so where they changed runners. When the runner ran through the main street in Kanab, the people ignored him. Didn't the 4,000 people in Kanab know that the Olympic torch bearer was running through their town!? I clapped loudly and a few people turned their heads.

I also liked to hike in Arches National Park. I stayed in motels in Moab and made day hikes on all the trails, including off the trails to where one can look up to Delicate Arch (shown on Utah license plates for years). Bill Miller and I made Jeep trips in Monument Valley (while it was allowed), Black Mesa, Canyon de Chelly, and White Mesa (see Section 2 on the Crab Nebula).

John liked the Capital Reef National Park area and bought property in nearby Torrey. He had an architect design a house for him and he planned to retire there. However, I convinced him that with his health problems, he should stay close to a hospital.

17
PACIFIC ISLANDS

I loved the Pacific atolls and went to as many of them in the 1960s as I could reach by plane. They are too widely dispersed to go by ship because that involved taking off too many weeks of vacation. I swam in the lagoons, laid in the sun, enjoyed the intense sunlight on the palm trees and plants, snorkel swam to see the rich coral life and fishes, and chatted with the local people who were laid back and friendly before catching on to the western lifestyle that involved working hard to acquire things. I liked their fruits and other foods. Sometimes with friends, I went to many islands in French Polynesia (Tahiti, Moorea, Bora Bora, Raiatea, Huahine, and Rongiroa). I met Craig Weinstein once in Hawaii while he was on a solo camping and hiking trip around the islands. He shared my love of those islands.

Craig learned that it was possible to hike across the island of Tahiti. That island is a round island 30 km in diameter with a smaller attachment to the southeast. It has central peaks up to 2400 m high. All the people live in villages around the rim; there are no roads into the interior. By previous arrangements, we employed a guide and two porters to take us across the island in four days. The interior was a wonderland of peaks, lush jungle vegetation, lakes, rivers, and streams. The guide and porter caught fish for our dinners. At one place the choice was to hike on a steep dangerous trail along the side of a lake, which the guide and porters did, or to paddle on our inflated air mattresses across the lake, which Craig and I did.

I became a friend of Andre Brooke, a French artist and architect who had moved to Tahiti as a young man. His paintings were mostly abstract; one of them is shown on a French Polynesian stamp. He was a friend of Conductor Kurt Mazur (1927-2015), so on one of my trips to Tahiti I brought him the recordings of the six numbered Tchaikovsky symphonies.

I often went to the Fiji islands. They, along with Bali, are favorite vacation destinations for Australians and have nice resorts. Most of my trips were to the main island of Viti Levu (120 km in diameter, peaks to 1300 m), but on one trip I also went to Venua Levu. Although most of the people on Viti Levu live around the rim of the island, some live in interior villages. There were no roads into the interior. I arranged to hike across the island in four days with a young man as a guide. Every afternoon on Fiji radio between 2 and 3 pm they broadcast messages to the interior villages, so the messages went out each day that two hikers were coming to a certain village and the inhabitants should be prepared to provide them living space and meals. The trail was easy except for the initial climb to higher elevations from Suva along the side of a steep hill. After that it was mostly rolling hills and meadows where the people raised livestock and farmed. The people spoke Fijian, but with translations by the guide, we got along well. They have a great sense of humor.

In May 1973 I went to Micronesia. The plane (a Boeing 727) stopped at Johnston Atoll; it is all military and we were not allowed off the plane. They had scraped all the sand in the atoll to make a rectangular island twice the width of the landing strip. Then we went to Kwajalein Island in the Marshall Islands which also belongs to the US Military. We were allowed off the plane but without cameras. However, the interesting island was Pohnpei, formerly called Ponape. It has high central peaks (790 m) and an area of 334 km^2. The runway was only dirt. The main village is Kolonia, which had 6,000 people.

The island has many lobsters and in the small Cliff Rainbow Hotel where I stayed ($10 per night, $10.50 for meals) they served lobster tails three meals a day. After several days my face swelled so much that I could hardly see out. A local doctor told me that it was due to an allergy to excessive iodine.

A fascinating ruin on the east side of the island is called Nan Madol. It is a set 92 ruins made of basalt columns brought with great difficulty from the western side of the island and laid crisscross. It had a complex pattern of enclosure, tombs, vaults, and houses. One can get to them only at high tides. They date from the Sau Deleur Dynasty of about 1,000 years ago.

I recommend the book *The Island of the Colorblind* by Oliver Sacks. To the east of Ponhpei is a low island called Pinelap. It was devastated several times by hurricanes in the past centuries. In one, all but 10 of the people died. One of those 10 was totally colorblind: no cones in his retinas. Such people have other strange characteristics, such as being extremely sensitive to light (they have to wear double sunglasses). The descendants of that person caused 10% of the people on Pinelap to be totally colorblind. This is different than the people (like me) who have color vision deficiency, which affects 8% of American men and 5% of American women. Such people have trouble telling blue from violet, red from green, or other confusions. The people on Pinelap finally wised up to the lethal effects of hurricanes and moved to the high island Pohnpei, where hurricanes do not kill all the people. But Oliver Sacks discussed the relation of colorblind people on Guam and the many people there having Lytico-Bodig syndrome and Parkinson's Disease.

From Pohnpei I was scheduled to fly to Truk Lagoon but there were strong cross winds at the landing strip. After two attempts the pilot said that if he tried a third time and failed, he would not have enough gas to make it to Guam. So we flew for an unscheduled flight to Guam, which I had never visited before. We were taken to the Guam Continental ($22 per night). It had 200 rooms in cinder-block buildings.

When I got to the hotel someone said "Helmut, what are you doing here?" It was the tourists Bob & Chris Taylor, whom I had seen earlier.

The next morning we flew to Truk (now called Chuuk). The lagoon is about 90 km in diameter with various islands around the perimeter and within it and a population of 40,000. The dirt runway was on Moen Island. The plane made a hard stop about 100 m from the end of the runway.

Truk Lagoon was Japan's main naval base in the South Pacific during WWII. Starting on 17 Feb. 1944 the Americans bombed the Japanese fleet for three days. Truk Lagoon now has the largest graveyard of ships in the world, including a submarine and a destroyer. One can scuba dive or ride in a glass-bottomed boat and see the many ships below. One ship has a mast that sticks above water and one can sit on. On a sunken cruiser one can sit on the 3-inch gun. Divers found that the ships contained aircraft, tanks, bulldozers, railroad cars, motorcycles, torpedoes, mines, bombs, munitions, and radios. There was a Seabee camp on Dublon Island of about 25 men who did civilian projects for the Trust Territory, such as building roads and water systems.

I stayed in the Truk Continental on Moen Island. I met two guys who came there for the snorkel swimming. They worked on Kwajalein Island. Tom Healey taught English and art. He told me that his father (Harold Healey) had been a painter on Kitt Peak and his mother (Freda) had been a cook there. His friend Larry Brantley worked for the government on Kwajalein as a computer expert on the Safeguard System. I also met Bob and Betty Newman, who worked for Shell at Brunei on north Borneo, installing a refinery control computer. I also met Tillio and Bruce Grillo from Eureka, CA, who sailed a 40-foot sailboat from California, stopping for weeks at Kwajalein and Pohnpei. I Also met Joe.... from Woleai Atoll in Yap. The atoll is 7 km wide, has a peak elevation of 2 meters, and a population of 1,000. Its main product is copra or coconut shells, but at the current price of $110 per ton or 5¢ per lb., it wasn't worth harvesting.

18

MARSHALL SPACE FLIGHT CENTER (1967)

I do not remember why, but I was asked to give talks on "Modern Stellar Astronomy" for four days in Sep. 1967 to engineers and technicians at the Marshall Space Flight Center in Huntsville, AL. The audience of about 100 were aerospace engineers, scientists and technicians, including many from Germany (Wernher Von Braun, etc.) who developed the V-2 rockets during WWII. I realized that I could not talk for ~six hours on each of four days, so I asked Peter Pesch to share in giving the talks. We were not introduced to the large anonymous audience; people came and went as their schedules permitted. Interesting experience.

19

EASTER ISLAND (1970)

Easter Island, or Rapa Nui, was discovered on Easter Sunday in 1722 by Jacob Roggeveen. It belongs to Chile and is about 3759 km (2331 mi) west of Santiago. From previous trips to CTIO I learned that the only transportation there is by Lan Chile's Boeing 707 from Santiago once a week. It flies for 5 hours from Santiago and stays at Easter Island a couple hours. Then it flies to Papeete, Tahiti and stays overnight. The next day it flies back. Thus one can stay on Easter Island for 2 hours, one day, or one week. I arranged to go to Easter Island in 1970. It turned out that the return flight was delayed a week because the 707 had to return to Hamburg for servicing, so I stayed on Easter Island for 15 days. At that time there was no hotel operating on Easter Island, but I could stay with a local family.

The Mataveri International Airport was built by the Americans in 1967 as an abort site for the space program. The runway is 3318 km (10,885 ft) long.

The "father" of Easter Island was Father Sebastian Englert (1888-1969), a German Caputian monk. He did everything that he could for the people, all of whom lived in the only village on the island – Hanga Roa. He learned the local Polynesian language (called Pascuense) and gave his sermons in that language. He helped the people in every way he could, numbered and mapped all of the statues (moai, nearly 1000), etc. Thinking that I would see him, what could I bring him because they had no TV and mail was infrequent? A ship came twice a year to bring supplies. When I was there they had run out of concrete so no construction was possible. Also it was out of flour, rice, salt, sugar, and wine. So I brought an encyclopedia year book that told what had happened in the world during the previous year. I also brought a portable tape recorder because the singing in the church was famous.

When I arrived on 14 October 1970 I was shocked to hear that Father Sebastian had died. He had gone to Washington, DC for a Smithsonian exhibit of Easter Island statues (moai). Like many other South Americans, he suffered from kidney disease from drinking too much wine. A Jesuit monk from Staten Island, NY named David Ready, who was a ham radio operator and had communicated with Easter Island frequently, had the sad job of telling the people there that their father had died (in New Orleans).

I was met at the airport by Alicia and Eugenio Hucke, my hosts. There was no terminal building but about 75 local people sold souvenirs on the tarmac. The Hucke's had a five-room house with no heating but electricity (220 v), a refrigerator, and cold running water. We agreed that I would pay them 130 Escudos ($6) a day for room and meals. We had a late dinner of rice and wieners. The Hucke's had three sons, one (20) in Chile, Eugenio (15) and Enrique (8). Eugenio Sr. worked on a sheep ranch in the interior of the island. We (including neighbors) talked until midnight, when I crashed. They spoke Pascuense and Spanish, with some English.

The next day I spoke with the Rapahango brothers; Bennito Rapahango was Alicia's brother. They would like to start a factory for canned pineapple that would be shipped on the semi-annual ships and they had the machinery, but lacked the concrete to build a factory. I gathered that Chile treats the Easter Islanders as dependents, rather than encourage them to be financially independent. All the major positions (e.g. postmaster, teachers, and nurses) are held by Chileans, enough though some of the local people were qualified to hold them.

Alica packed me a huge lunch and I hiked along the west coast and noticed that each moai was different (Fig. 19-1), as though modeled for a different person. All were knocked down in the civil wars although Heyerdahl recently stood up some of them up on platforms. Dr. William Mulloy from the Univ. Wyoming spends half-years on Easter Island on a UNESCO grant. I talked with two men whom he employed full time.

Fig. 19-1. Some of the 1000 statues (moai) on Easter Island. These were finished at the quarry and were waiting to be moved (by "walking") to the final sites.

On Saturday Bennito Rapahango and his wife took me to Anakena on the north coast for a barbeque, together with men from the air base. We went swimming in the ocean, where the water temperature was ~70F deg.. The men played a game like horseshoes, throwing steel pucks ~30 ft. The barbeque was meat with wiener rolls, cold vegetables, and Chilean wine. The men played a game with Spanish cards (numbered 1-7, J-K). Two snorkel swimmers brought in two pufferfish, etc. We returned to the Honga Roa at 2:30.

In the afternoon I climbed (1.5 hr) Rano Kau, the crater at the SW corner of the triangular island. Beautiful views of the lake and reeds in the bottom of the crater. The rim is the location of Orongo with houses under the rocks and many petroglyphs cut into the rocks. More about that below. In the evening, I went with Eugenio and Enrique to the theater near the church in the village to see *Boy on a Dolphin* with Spanish subtitles. The theater had a dirt floor and benches. There were about 200 people. The sound was very poor. There were breaks when they changed reels. Cost 5 Escudos (20 cents) per person. The parking lot had one Jeep and one tractor.

The next day was a rainy Sunday. After a breakfast of two fried eggs on meat, I went to the Catholic mass at 10, one of three masses. The interior of the church was rectangular with idols and pictures on the walls. The two priests talked in Latin and Spanish. The single men sat on the left, the single women on the right. Much standing, sitting and kneeling. The songs were started impromptu by a lady in the front. They were mostly Tahitian songs, some Rapa Nui, sung in Spanish. The service took ~45 min.

In the afternoon Enrique, Jr. took me to the Ana Kai Tangata, a cove near the village. Colored paintings in the cave. Spectacular waves and surf. Easter Island has no coral reef around it and no good harbor. Lunch consisted of fresh tuna and avocado, soup, and ice cream (made on the island). I showed Eugenio Jr. some of my spectra from CTIO and tried to explain what they were. I walked to the north end of the village and met the pastor of the Pentecostal church, one of the two churches in the village. I walked past the new hotel that was being built, a prefabricated building too far from the coast to see the spectacular surf. Scattered rain throughout the day.

I should explain that the Americans built an airstrip between crater Rano Kau and the village of Honga Roa, the only inhabited village on the Island. The airstrip (3318 m = 10,885 ft.) was originally built in 1967 as an abort site for the space program. At the time the American Air Force was monitoring the fallout from the French nuclear tests in the Society Islands. The Americans also drilled a water well in the side of Rano Kau and built a purified water system for the village. At the time that I was there, the Americans were withdrawing from the island because the Communistic government of Allende was being voted in in Chile. Military planes (C-141s) came in to remove their equipment. The Americans left the airstrip and water system to the islanders.

The island became overpopulated in the 18th century with about 4000 people. As a result, all the trees were cut down for firewood and warfare for food broke out between tribes. Some of the people lived underground. Hence the strange birdman cult events, described below, centered on Orango. Chile captured many men and forced them to work in the guano (bat droppings) fields on the islands off the Peruvian coast until international pressure in the 1870s made them stop that and sent the men back to the islands. They brought European diseases to the island that reduced the population to 123. Hence knowledge of the past, known by the old people, was lost, such as how to read the rongorongo boards. The current population is about 7800. Almost all of the coastal area is in the Rapa Nui National Park with the central part left for the people to graze animals and plant crops.

One day I hiked to Maunga Tere Vaka (507 m), the crater on the north side of the island. Beautiful view of the whole island. On the way back, I talked with two men who wanted to trade wooden carvings for my clothes. Because the island has no mature trees, the locals have to get wood (using a motorboat) to an uninhabited island, which has trees, about 40 km to the northeast. The carvings I got were made by islanders and cost me $14-16 each, but a friend who visited the island recently said that now all the carvings for sale came from China! To get one carved by an islanders, he was told to go to the local jail. I bought a carving from Mio Tahiti wood of a "birdman" – upper part like a bird and lower part like a naked half-starved man. But I traded my tape recorder for a copy of a rongorongo board (15 inches long). It contains the only written language in the Pacific islands. The missionaries convinced the islanders to destroy most of the original ones because it was a "heathen" language, so now there are only 15-20 originals left in the world – not enough to be able to decipher the language.

According to tradition, when King Hotu Matua landed on the island about a thousand years ago, he brought a language written on banana leaves. They were disintegrating, so be had them transcribed onto hibiscus wood, using shark teeth as tools. Only the royalty could read them. Reading the picture-like text involved reading one line (with no punctuation) and then turning them around to read the next line, or two men facing each other reading alternate lines.

When overpopulation caused warfare between the tribes, they settled for the control of the island on the birdman culture. Every spring the sooty terns would come past and lay eggs on small offshore islands called Motu Iti (Fig. 19-2). A representative of each tribe would watch from the sacred site of Orango at the top of crater Rano Kau for the return of the sooty terns. Then each representative would climb down the steep 1000 ft. cliff below Orango, swim through the shark-infested water to Motu Iti, put an egg in his mouth, swim back and climb up the cliff. For the first man to return with an egg, his tribe ruled the island for one year. The chief of the tribe became so sacred that he had to spend the year in a cave. While the people waited at Orango for the birds to come, they carved hundreds of drawings in the basalt rocks and slept under rocks. Orango is a fantastic place to see because of its high precarious location and the many drawings.

I climbed down into the crater on the south side, ate guavas that grew there, and tried walking on the totoro reeds that covered most of the lake. Then I hiked down to the village, passing the American ionosphere physics station.

My next hike was to Rano Raraku near the east end of the island. That is the quarry where the moai were carved. It has a small lake with marshy edges on the west side where the sheep and other animals drink. But the site is of a volcanic tuft that is easy to carve. The moai started out on their backs as the fronts were shaped, leaving a ridge along the back for support. They were carved to have breechcloths. After the fronts are finished, the ridges were cut and the statues are stood up outside the quarry to finish their backs. There were a dozen statues outside (Fig. 19-1), waiting for transportation to their final sites; they got buried up to their hips from the recent erosion of dirt.

Fig. 19-2. The small islands (Motu Iti) off the coast of Rano Kau where the sooty terns came each spring to lay eggs.

One statue was inside the quarry and partly finished. To cut the basalt, they used harder rocks; I don't know where they got them from. I picked up one of those hard pointed rocks and spent about 5 min. hitting into the basalt. I got about 3/8 inch in, so one can extrapolate that a half dozen men could carve a statue in a couple months.

A long-standing question is how the statues were moved to their final sites several miles away. They were probably not pulled on their backs because the backs and fronts were sculpted and would be damaged. The local people said that "they walked". What they meant was that the statues were stood up vertically with ropes around them held by a couple dozen men. They tilted the statues to the right and pulled the left sides forward a foot. Then they tilted the statues to the left and pulled the right side forward. In that way they walked the statues along a horizontal path to the places where they were stood up on platforms.

I became acquainted with Father David Ready, the new priest in the village. Actually, after Father Sebastian died, the Catholic Church selected a German priest, who was a disaster because he made no attempt to communicate with the people – just gave his sermons in Latin. Father Ready gradually replaced him. He told me about Father Sebastian's death. Father Ready was so new to Easter Island (about six months) that he had not even been to Orango. I took him up there (about a 70 min. walk) and explained to him the birdman ritual. He was impressed. To date he was more concerned with the financial and social welfare of the people in the village. I gave him my copy of Father Sebastian's book: *The Island at the End of the World* (Realton Books, 2005).

I talked with Bill Clay, the American coordinator for the new hotel being built by the Chilean government at a cost of $3M with almost no islanders being employed during the construction. I heard about Allende's election and later that the head of the army, General Rene Schneider, had been assassinated on 24 Oct. The local people didn't like Allende because they suspected that he would drive out the USAF and American tourists.

I gave some of my clothes, kitchen utensils, and sewing materials to Alicia and to Juan Haoa, who gave me two carvings.

One day I hiked all the way to Maunga Vai Heva crater (370 m high) at the east end of the island. While I was napping after Alicia's huge lunch, a C-141 flew a hundred ft. above me to land at Mataveri airstrip. That day I hiked 11.7 hours (6:45 am – 6:30 pm), and about 31 miles. My estimated total hiking during two weeks was 175 miles.

The last full day I hiked up Rano Kau again. I gave my hiking boots to Herman Otu and Juan gave me a carving (I do not know what happened to all those carvings because I have only three left). I gave my jacket and other things to Alicia and my canteen and cartridge belt to Bennito. He was earning $58 per month. He said that the Americans were leaving in December (I was there Oct. 15-30, 1970). The Lan Chile plane had mechanical difficulties and stayed overnight. I was uncertain about the reliability of the plane and what was happening in Chile, so I decided to fly to Tahiti and then directly home.

In 2009 a friend (Wiphu Rujopakarn from Thailand) visited Easter Island and found Eugenio Jr., who was still living on the island. He was married and had two sons who were visiting the island for the holidays. One was an architect and the other in biochemistry in Chile. Wiphu showed me pictures of Eugenio, Jr., aged 58, and one son. We then communicated by email, but it is difficult to communicate with someone from 50 years ago and having such a different environment.

20

THE GALAPAGOS ISLANDS [1971]

I have often wanted to visit the Galapagos Islands, particularly after reading *The Voyage of the Beagle* by Charles Darwin. I signed up with Lindblad Tours to take a 10-passenger ship (the *Golden Cachalot*) from Guayaquil, Ecuador to the Galapagos for two weeks. Air fare to Ecuador was $572 round trip and the tour was $1450. On January 30-31 I flew to LA to Mexico City to Panama to Quayaquil on the Pacific coast. A man from Metropolitan Touring took me to Hotel Atahualpa ($14.12 a night, paid by the tour).

Actually we had been told on January 28 that they were having trouble with the ship's engine and two couples had dropped out. The remaining six were:

- Dr. Spinks & Cachs Marsh, radiologist from W. Hartford, CN,
- Dr. Charles Field, anesthesiologist in the Mayo Clinic in
- Rochester, MN,
- Ms. Joan, social worker from Morristown, NJ,
- Ms. Ruth Pennebaker from New Orleans and me.

We were taken to local sites (e.g. a cocoa plantation) and wondered around town. Then David Balfour of Metropolitan told us that the bearings on the ship failed again and trip was cancelled. Other options were offered. All six of us opted to go on the *Lina A*, a 55-passenger ship, to major islands in the Galapagos for one week. Four of them then chose to go to Machu Picchu the second week but Charlie and I decided to spend the second week on Santa Cruz, the headquarters town for the Galapagos, and also on a small sailboat, visiting smaller islands. Lindblad refunded $486 to me.

On 2 Feb. we took a plane (Tame 191) for the three-hour flight on a DC-6 with 43 other passengers to Baltra Island, just north of Santa Cruz. Baltra is flat and barren, and the terminal was a two-room building. The landing strip was built during WWII by the Americans to protect the west coast of S. America from the Japanese. The Galapagos Islands, straddle the Equator, and are on Central Standard Time; Ecuador is on Eastern Time. We took a bus to the *Lina A*. Charlie and I had one of the two cabins on the top deck. Very nice! We also had all of our meals at the Captain's table.

We boarded the *Lina A* and sailed past Daphne Minor with its steep sides to Sulivan Bay on Santiago and nearby tiny Bartholomew Island, named by Darwin for Bartholomew James Sulivan, who was a lieutenant on the HMS Beagle. We went swimming off the beach and saw sea lions, sally lightfoots, crabs, etc. Dinner was local lobster.

The next day we sailed around the northern tip of Isabella, the largest of the islands and it has with five craters, to Tagos Cove. We saw lots of penguins whose ancestors apparently swam north along the Humboldt Current and decided that they liked the Galapagos better than the Antarctic. They are smaller than the Emperor penguins of the Antarctic. The blue-footed boobies have very vivid blue feet. We also saw flightless cormorants. We landed on the shore of Fernandino, whose central crater (Narborough, 1500 m = 4900 ft) is still active. The island was named after King Ferdinand, who sponsored the voyages of Columbus. On 13 May 2005 it started a new eruption and its central lake dropped 1000 ft. Please see the lead article in the 31 July 1970 issue of *Science*. The lava beds around the crater are young and extremely rough. I went swimming with the sea lions and the young ones could swim around me faster than I could turn. They seemed to enjoy playing with tourists. There were also white boobies (with yellow bills), hawks, green herons, marine iguanas, and Darwin finches. After dinner we had a talk and slides, as usual. Isabela has the largest number of Galapagos tortoises; a lesser number is on Santa Cruz Island.

The next day we went to Genovesa (Tower) Island, named for Genoa, Italy. It looks like a partially submerged crater with Darwin Bay providing a safe harbor, but dangerous to enter because of the submerged arms of the crater. The surface area is 14 sq. km. It is also called "bird Island" because of the many species of birds there. They include:

- Red-footed boobies that build their nest in trees;
- Masked boobies that are mostly white with red feet and black around their beaks;
- Yellow-crested night herons;
- Lava gulls, brown and grey;
- Swallow-tailed gulls, grey and white with orange eye rings;
- Frigate birds that are fed in their nests for 18 months; the males have large red pouches under their beaks to attract the females; they can fly and sleep with the pouches inflated;
- Mocking birds, grey & white, very tame; they will drink water out of a cup held in people's hands;
- Also present are finches, which attracted Darwin because they differ on different islands. Only three species may now be distinct; actually many other animals and birds differ on different island, mostly between tall and low islands;
- Blue-footed boobies which generally lay only one egg yearly; if it lays two, the first is preferred for food and the second survives only if the first doesn't;
- Marine Iguanas, which are smaller and black on Tower Island. Charlie and I went swimming for 10 minutes. After boarding, the Captain warned us about sharks around. After crossing the Equator a fourth time, a crew member played King Neptune

and made everyone get on their knees and kiss a dead fish. In the afternoon we went to Black Turtle Cove on Santa Cruz, where we saw hundreds of green turtles, about 10 manta rays, oysters on the mangrove branches, and sharks.

The next day we went to Baltra Harbor were we went swimming after borrowing the Captain's mask and snorkel. We saw jewel fish, black with bright blue spots. We met Edward "Tosh" McIntosh, who made arrangements for Charlie and I for the second week of the trip. I remember him to be a biology Ph.D. student from the Netherlands. He recommended a three-day trip on burros to the Santa Cruz highlands with a guide and John Fitter, and then a four-day sail with Gusch Angermeyer to Barrington Island, Plaza, Daphne, Bainbridge, Sombrero Chino, etc. Sounded great! The other passengers disembarked and the new group (Philadelphia birdwatchers) boarded.

Tosh took Charlie and I to North Seymour Island, known for its large population of blue-footed boobies. We saw their mating procedures. The male exhibits his red pouch and the female responds with giving him a stick (part of the nest-building ritual). They clash beaks; the male scares other males away. After the eggs are hatched, either parent will guard the babies.

During the night the *Lina A* sailed to Hood (Espanola) Island, named for Viscount Samuel Hood. Its marine iguanas change color during the mating season. It is the only island having the waved albatross and they take off from the steep cliffs, going to places near Ecuador and Peru.

Then we sailed to Point Cormorant on Floreana Island. We swam with the sea lions, jewel fish, Moorish Idols, etc. and on shore to see a dozen flamingos. Then to Post Office Bay. There is a mail box (originally it was a barrel) there; 19th century whalers dropped off and picked up mail there. Now the crew of the *Lina A* picks up the mail monthly and gives it to any ship sailing to Australia or New Zealand. I think that my postcards actually got to the States.

The next island visited was Santiago (San Salvador or James). It has a shallow saltwater lagoons and the buildings around it are left from when the lagoon was mined for salt. The buildings were not there in 1684 but mapped by Darwin in 1835. However, there were jars of Spanish quince marmalade from Cowley's *Bachelor's Delight* of 1684 found there.

During the night we navigated to Academy Bay on Santa Cruz and stayed at the only hotel there at that time. The owner (John Marlena Nelson, an Ecuadorian) was out in the field with a crew from the Encyclopedia Britannica filming four films for TV. They were the only other customers, besides Charlie and I, at the hotel.

We went first to the Charles Darwin Research Station where Assistant Director Dr. T. De Vries showed us around. (I do not know whether he was related to Dr. Hugo de Vries, the Dutch botanist who named and postulated evolutionary mutation). Dr. De Vries had been there for 5.5 years and was returning to the Netherlands. The station hosts visiting scientists (45 in 5 years) and does research on local animals. For instance, the staff found that there were no young tortoises because the islands have been infested by rats from ships; the rats ate the tortoise eggs and young. Therefore, they have gathered up 300 eggs each season and nurtured them at the station. After 10 years the tortoises were large enough (one foot long) to no longer be vulnerable to the rats. That year they had returned dozens of tortoises to their home islands and they were surviving. There were only seven saddle-back tortoises left on Hood Island, so the staff brought to the station one male and three females. The first batch of eggs were just hatching.

The Charles Darwin Research Station had an annual budget of $24,000 and was sponsored by the Smithsonian Institution and other environmental groups. The Director was Dr. Rolf Sievers from Germany. Our group of six from the ship walked back to the village (Puerto Ayora) and they consumed a quart of beer each. They then left us.

A local resident was Julian Fitter, who was a naturalist, writer, environmentalist from Oxford. His book, *Wildlife of the Galapagos*, is still one of the best. When I was there his wife, Mary, was pregnant and he was uncertain whether the pregnancy was proceeding normally. When he heard that there were two doctors visiting the island (Charlie and Don Boots, Sr. – see below), he was eager to consult them. He told the two doctors of the tests that the local doctor had done and their outcomes. Charlie and Don agreed that the local doctor was doing the right tests and the results were normal. Later I learned that Julian had taken Mary to Quito for the birth and it went well. I corresponded with Julian's first son, John, for several years because he liked to collect stamps. I later sent him a copy of *Scott's Catalog*.

Don Boots was an interesting person with whom I corresponded and met for decades. He was a doctor in Santa Barbara who had a sailboat (40 ft., I believe) named the *Spencerian* and who decided, with his family, to sail around the world. He sold his practice and they sold their house. He and wife Catherine (called Cassie) had four sons. The oldest was in the university and chose to stay there, but the three youngest sailed with them. They took correspondence classes. The oldest of the three (Randy) eventually dropped off in New Zealand, became a fisherman, and married there. The two youngest were Don Jr. (17) and David (14). Both were typical blond American kids who brought their skate boards, probably the first, to the Galapagos and Tahiti. Sailing across the ocean is not easy because someone aboard has to be alert 24 hours a day for fear of colliding with ships. Therefore after a week or two of doing that, when they reach a port they tend to stay for a month or two to recover. On later trips I saw the Boots family again in Tahiti and in Tucson. Don Jr. and John Fitter joined us in the trip up into the highlands of Santa Cruz, where the tortoises go to mate.

The highlands of Santa Cruz Island are very rich in diverse vegetation. One can find there bananas, plantain, lemons, oranges, grapefruit, pineapples, papayas, mangoes, chirimoyas, guavas, watermelons, coconuts, plums, figs, medlars, tamarinds, acerolas, potatoes, camotes, otoya, yucca, avocados, breadfruit, tomatoes, leeks, peppers, parsley, pumpkins, lettuce, coffee, sugar cane, maize, tobacco, and balsa trees. In fact, the Galapagos still have 95% of all the flora and fauna that existed there before men came.

A man with one horse and two mules (Don walked) arrived at 3 pm and we stayed overnight at the Horneman's house. The Horneman's had moved their house from Norway in 1927, where they bought it for $10. They lived in the Galapagos for decades, but when the husband died in 1969, Mrs. Horneman went to live with their daughter, who lived in Norway just north of the Arctic Circle. Imagine moving from the Equator to north of the Arctic Circle! The house looked like she had left the day before, with the kitchen, library and bedrooms intact. It was kept in shape by an Ecuadorian couple with three children and many dogs and cats. There was a lunar eclipse at midnight.

Steve Devine arrived in the evening and we went with him the next morning to his horse ranch at the top of the highlands, where the tortoises spent part of each year. When he learned that I was from Tucson, he asked me to call Dr. Robert Hastings, Bruce Jacobs, and John Steiger when I returned. Jumping ahead, I did and learned the following. Steve came from a rich Chicago family and was attending the Univ. Arizona on the skimpy allowance of $7,000 a month during the 1930s depression! He broke his leg and Dr. Hastings Sr. set it. Steve did not have the money to pay for it, so he gave Dr. Hastings the deed to his property south of Tucson. In 1971 I talked to Dr. Hastings Jr. who was surprised and shocked because the Hastings never heard from Steve Devine since the 1930s when he left Tucson. They had heard that he went to the coast and boarded a ship, but nothing after that. I asked about the property Steve gave to his dad. He did not own it, but had a lease with the option to buy it. But Dr. Hastings Sr. was drafted as a doctor during WWII and with a small salary, he could not afford to pay the payments. I asked where the land was. Dr. Hastings Jr. said it was the land now occupied by Davis-Monthan Air

Force Base, Raytheon Missile Co., and Tucson International Airport, land worth billions! I wrote Steve that, but I doubt that he cared and he did not reply.

Steve's ranch was one section in area with 34 pastures. Most of the tortoises have numbers; we saw #187, 106, 109, and an unnumbered one, but the ponds were dry so most of the tortoises had left. Steve's "house" was simply a lean-to. We had a good dinner of chicken, went to bed early, and got up at 5 am. Steve regularly charged $35 for use of a horse, lodging, and meals. We dropped off the horses and mules at Bella Vista and walked to Puerto Ayora. It was hot and humid walking so we enjoyed showers and many Pepsi's at the hotel.

On the way down we saw three ships off shore. One was the HMY Britannia, the royal ship with a crew of 270, an Ecuadorian naval ship with the Ecuadorian minister of something aboard, and a tender. The Britannia (commissioned 1954, three masts) had navigated there to meet the royalty, who flew from England. They consisted of Prince Philip, Lord Mountbatten, Princess Alexandria, etc. For those like me who are confused by British royalty:

- Prince Philip, Duke of Edinburgh, husband of Queen Elizabeth II, born in 1921 as Prince of Denmark and Greece; since 2017 he has withdrawn from all royal responsibilities.
- Lord Mountbatten, born 1900; during WWII he was the Supreme Commander of the Asian sector.
- Princess Alexandria, Lady Ogilvy, born 1936, first cousin of Queen Elizabeth II.

After the four of us (Charlie, Don, John, and I were cleaned up from our hike, we waited outside the hotel for lunch. We were told that we would be served after the royalty, but we figured that the left-overs would be good (they were: local lobster). The royalty was waiting too, but only Prince Philip come over and talked with us. We were in short pants, of course, and the boys were shirtless. Philip asked if we were locals, so I introduced us. He asked if I was doing astronomy there, and I said "No, just visiting". Charlie said that he was an anesthesiologist, using the British pronunciation, to which Philip chuckled. John said that his father (Julian) had met Prince Philip in Oxford.

Prince Philip was the titular head of the British organization that helps sponsor the Darwin Research Center, so he gave walkie-talkies and binoculars to the staff and 48 soccer balls to the school.

Then we went to the Boots' *Spencerian* for pineapple juice with Don Sr. and Cassie. I gave him my copy of *The Journal of Jacob Roggeveen*, the voyage in 1722 when he discovered Easter Island on Easter Day; hence the name he gave to the island. I gave John Fitter my copy of *The Fatal Impact* by Alan Moorhead, who described the effect of Europeans during 1767-1840 on Tahiti, Australia, and the Antarctic. We decided to go snorkel swimming but it was too late to go to the "Crack", so we went to the beach. Prince Philip's party was just leaving. We swam until someone spotted a shark.

We met Gusch Angermeyer, who would take us during the next four days to smaller islands aboard his sailboat. The Angermeyer family of the parents and four boys (Carl, Gusch, Hans, and Fritz) left Germany when the Nazis were taking over and ended up living in the Galapagos. Gusch has a son Franklin and Fritz' son is Fiddi. They set up the first touring businesses with sailboats. Carl built his house on a point in Puerto Ayora that was inhabited by marine iguanas. After the construction was finished, the iguanas moved right back and regularly went in and out of his house. Gusch built *Gusch's Cave* with all kinds of interesting things (shells, fossils, starfish, etc.) that he had found. Gusch's son Franklin, who went with us, had been partly drowned once, which left his brain injured.

On February 12 the *Britannia* left at 6 am for Easter Island and Pitcairn. The Ecuadorian navy ships had left. The royal party had not seen the poor houses in the village. We (Gusch, Franklin. Charlie, and I) did not leave until 11:30 because Gusch had trouble with the batteries on the *Liv* (Norwegian for Life), his 25 ft. sailboat. It is cat-rigged and sleeps four. It took us four hours to sail to Barrington (Santa Fe) Island, between Santa Cruz and San Cristobal Islands. It is said to be the oldest of the Galapagos Islands at 4.5 million years. It is a low unpopulated small island (24 sq. km) with many oputia cacti. It has species that are different than those on other islands, such as the leaf-toed geckos, the Santa Fe marine iguanas, and the Santa Fe land iguanas. In the lagoon, Gusch caught a gropper (baccalao), which he fried for dinner.

The next morning we used the dinghy to land and we hiked inland to a place where reportedly the ground is hot. We didn't find it. He hypnotized a land iguana so that he could pick it up – until he let it wake up. We went swimming in the lagoon. In the afternoon we sailed for three hours to the Plaza Islands (named for a former President of Ecuador), located just east of Santa Cruz Island. It had many sea lions, who were very noisy. There were sharks and blower fish around the islands. The phosphorescent plankton in the water was beautiful in the evening. Gusch caught a mullet for supper. Wonderful scenery on the islands with swallow-tailed gulls, sally lightfoots, a red iceplant, rocks with green moss and white and red lichen. Gusch fed bananas out of his hand to the land iguanas; they even nibbled on his toes.

The next day we sailed to the channel between Baltra and Santa Cruz. Gusch dove for lobster and brought up four for dinner. I saw one 6-foot shark and four 2-foot ones. There was phosphorescent plankton again in the evening. After that we went to Daphne Major, which is nearly inaccessible due to cliff overhang, but Gusch found a break and shouted instructions to us. Inside the crater there were hundreds of blue-footed boobies in all stages of courting, nursing, hatching and feeding. We walked among them, but I refrained from getting within five feet of them.

We sailed to the coves on the west side of Santa Cruz but the swell kept us from mooring. Instead, we sailed to Baltra where Lt. Mario Nieto, head of the Ecuadorian Air Force Base on Baltra, invited us for drinks, and we invited him on board for dinner. Beautiful luminous plankton again.

The *Lina A* had arrived and Kevin McDonnell of Lindblad Tours met us. We showered and shaved in his cabin. Dr. De Vries was there on his way back to the Netherlands. I gave Gusch my copy of *Morphological Differentiation and Adaption in the Galapagos Finches* by R. I Bowman. I promised to write him, and I did.

The next morning we flew on TAMI (Ecuadorian military air transport) to Guayaquil and returned to David Balfour the flippers, mask, snorkels, and list of our expenses. I bought 6 lbs. of Ecuadorian coffee (strong stuff). Had dinner with Charlie and Dr. DeVries. On Feb. 17th we were taken to the airport and flown to Mexico City and LA. The immigration official in LA remembered me. Then I took a flight to Tucson.

I had an extensive correspondence with Charlie, Gusch, John Fitter, and even with Dr. Tim Simkin of the Smithsonian who was the co-author of the *Science* paper on the 1968 collapse of the caldera Fernandina. Charlie wrote that when he told people that he had visited the Galapagos, most of them did not know what they were, why they were important, and even in which ocean they are in. He said that he could tell which people read and which didn't. Now, of course, there have been many TV shows about the Galapagos, so many people who do not read recognize them.

My favorite book about the Galapagos is the two-volume *Galapagos: The Flow of Wildness* by Eliot Porter (1968).

21

GEORGE VAN BIESBROECK PRIZE

George Van Biesbroeck (1880-1974) was a Belgian engineer with a strong interest in astronomy. He started work on roads and bridges, but by age 24 he started work at the Royal Observatory of Belgium at Uccle. Then he got a degree in astronomy at Gent Univ. and worked at Heidelberg and Potsdam Observatories. In 1915 he was invited to work at Yerkes Obs. With his family, they took the dangerous wartime voyage across the Atlantic. At Yerkes he determined positions of double stars, comets, asteroids, and other solar system objects. He did many useful things for Yerkes, such as finding the site for the McDonald Observatory and measuring the gravitational attraction to light during eclipses, as predicted by Einstein. He observed until a month before his 94th birthday, having published 673 papers. After his death several of us in Tucson saw his final measures through publication. Carl D. Vesely, a high school chemistry teacher in Tucson, and Dr. Brian Marsden of the US Naval Observatory published four papers in the AJ of Van B's observations of comets and other solar system objects. I collected from his notebooks and published 2200 of his observations of 700 visual double stars.

A local committee (Tom Gehrels, Pat Roemer, Uwe Fink, one other person, and myself) obtained donations to pay for the publication of his last papers. There was money left over. Because most of Dr. Van B's measures and work were done to help others, we decided to give an annual award to young people who have done unselfish work for others in astronomy. Sixteen awards were made in 1979-1996.

Among the prizes previously given by the AAS, all were made for personal outstanding research; none for unselfish contributions to the field. Therefore we felt that the Van Biesbroeck Prize should be a national award administered by the AAS. However, the AAS Council demanded that the award be by a non-profit corporation so that others could make tax-free contributions. I worked with a local lawyer, paying him $5000, to incorporate the prize. When that money was used up, he contributed pro bono work. In 1997 the prize was turned over to the AAS and it has been given to 18 astronomers since.

The first recipient was announced in 1997 at an AAS Council meeting as me. I immediately screamed "That's not legal! The rules of both the original Van B. committee and the new AAS committee state that members of the award committee were not eligible to receive the prize". They replied that they had consulted their lawyer; that there were four days between the termination of the Tucson committee and the start of the AAS committee, and during those four days they awarded the prize to me. I doubt that no one ever worked so hard to get a prize for themselves.

22

THE ASTROPHYSICAL JOURNAL

In the 1960s before the availability of the internet, we tended to read the major journals and tried to remember the papers in our fields. At that time there were only a half dozen astronomical journals in English that were published (mostly monthly) plus many more observatory publication that came at irregular times. The journals had annual indexes, at least. But if we searched for an older paper (or changed fields), it was difficult to find it.

The Astrophysical Journal (ApJ) was an exception in that it had General Indexes for every 25 volumes (12.5 years) in addition to the Annual Indexes. I used the General Indexes heavily because I changed fields of research often. But when an index for volumes 101-125 was due, nothing came. I waited a year and then phoned Chandrasekhar and asked him about it. He said that he did not have time to compile one (the author indexes could be compiled by a non-astronomer but the subject headings had to be chosen by an astronomer). I asked him if I compiled a manuscript, would the Univ. of Chicago Press published it. He said yes.

I was prepared for that. I had hired a secretary (Eleanor S. Biggs) to organize the spectroscopic records from Kitt Peak and maintain a card catalog of all the spectra of each of the stars observed there. After it was started, she had half time for other things. She had a taste for careful numerical work if she was not rushed. So she and I compiled a manuscript for the ApJ General Index for Volumes 101-135 (1954-1962), which was published and distributed free to all subscribers. Almost immediately we started the next General Index for Volumes 136-145 (for years 1962-1966). After that Chandra asked me for another, for Volumes 146-165 (1966-1971).

In November 1970 I received a phone call from Chandra, asking whether I was coming to or through Chicago anytime soon. If so, he said that he would meet me at the airport. I replied that I was coming to Chicago next month for an AAAS Council meeting and that I would be pleased to visit him in his office. It turned out that my mother and I visited him in his lakeside apartment. He offered to make me the next Managing Editor of the ApJ. I replied that I did not know as much about astronomy as he did and did not feel qualified. He said that the main problem for the next editor would be to organize the operation of the ApJ, which was growing rapidly. He was working with one Associate Managing Editor, one secretary, one copy editor and one production manager, but couldn't keep up with the flow of manuscripts. In 1970 the ApJ published 630 papers. When I finished after 1999 we published about 2500 papers a year with about 40 Scientific Editors and production people.

I would have done almost anything that Chandra requested, so I agreed. We overlapped for several months in mid-1971 and I acted myself from October 1971. I retained Dimitri Mihalas as Associate Managing Editor and Donald Osterbrock as Letters Editor. When Osterbrock went to the Lick Observatory to become its Director, he was replaced by Alex Dalgarno in 1973 who continued as such for the duration of my editorship and part of that of the next two Editors. Later my title was changed from Managing Editor to Editor-in-Chief. In the late 1970s Stephen E. Strom replaced Dimitri as Associate Managing Editor.

All previous Managing Editors of the ApJ had been members of the Univ. Chicago's Astronomy Dept., which was at the Yerkes Observatory. By selecting someone outside the Univ., the Univ. may have been nervous because it did not have control of me. More importantly, Chandra felt that the primary American astrophysical journal should belong to the American Astronomical Society (AAS). For one thing, he worried that the Univ. Chicago Press might look upon the ApJ as a source of profit, rather than as a non-profit journal. That proved to be a correct prediction because years later (after the ApJ was not longer published by the Univ. Chicago Press), the Press looked upon its 40-some journals as a sources of profit.

Therefore Chandra and AAS President Martin Schwarzschild negotiated a transfer of ownership. There were discussions about how much of the ApJ's reserve fund should be transferred. Also they agreed that the finances of the ApJ and AAS should be kept separate. That led to a problem in the late 1970s. The AAS realized that it was costing the AAS expenses for the bookkeeping and funding of travel expenses for the Publications Board. It seemed logical to charge the ApJ for those expenses, but a transfer of funds between the ApJ and AAS was in violation of the ownership agreement. I was worried that if the AAS took 3% of the ApJ income, it could grow to 4, 6 or 10% at the whim of future AAS Councils. But I lost that argument and I frankly do not know if the original percentage has increased.

Chandrasekhar continued his spectacular research career and received the Nobel Prize in Physics in 1983 with William A. Fowler. It was for Chandra's discovery of the maximum mass of 1.4 solar masses for white dwarfs and Fowler's understanding of the evolution of stars (with Fred Hoyle and the Burbidge's). Chandra would work on a different part of astronomy every 5-10 years, publish many papers, and then summarize the results in a definitive monograph.

I decided that I would make no major changes to the ApJ during my first year until I understood why things were done as they were. The sole exception concerned the font size for the abstracts. They were printed with smaller letters than for the main text, but as the journal grew beyond the abilities of most readers to read all of the papers, we relied more on reading the abstracts to determine whether we would read the entire papers. So why print them with smaller font? I asked the Press staff why that was done, and they replied that it was conventional and could be changed. So I asked them to use the same letter size as for the text, and they did that.

After the first year I started to make a series of changes and improvements as listed below.

1. In 1972 I addressed the complaint of always having the halftones at the ends of issues, meaning that readers had to flip back and forth between the text and illustrations. That was necessary because to place the half-tones on glossy stock within the papers meant that in each paper copy a press person had to hand cut the signature and insert the illustrations. But with the improvements in paper stock, not all illustration needed to be on glossy stock. I published illustrations of three figures of different kinds on glossy and text stock to show that some could be placed on text stock and retain all the details.
2. Also in 1972 we announced that all papers in the ApJ will have printed on them the subject headings that will be used for them in the Annual and General Indexes. The initial choices, from a list of about 200, would be made by the Managing and Letters Editors. During the reviewing and copyediting phases, the authors could add or subtract from those choices. The idea is that readers will become familiar with where to search in the Subject Headings for that and similar papers. Also that allowed the Press to generate Annual and General Indexes without additional work by an astronomer.

    We announced a lower page charge ($10) for planographed pages. We announced that the Journal would publish color reproductions, although they were expensive at that time. The Press selected a new compositor who was cheaper.
3. The Supplements became so popular that we decided to have them come out as monthly issues, rather than singularly.

Dr. Eugene Garfield of the Institute for Scientific Information said that the Supplements received twice as many citations as Journal papers on the average.

4. In 1973 after experiencing a 75% increase in two years in journal content, we either had to increase rates or find ways to publish the ApJ cheaper, or both. To save money, we went out for bids and moved the composition for Part 1 to the Institute of Physics (IOP) in England and for Part 2 to Worchester, MA. We bought paper directly by the train car loads. We reduced some of the figure sizes. We also increased the journal page sizes from 6.5x9.5 to 8x10.5 inches (20.3x26.7 cm). (In 1984 it was increased to 8.5x11.25 inches or 21x28.6 cm). We also asked authors to write more concisely. We moved some papers to the Supplements, especially those containing much digital material. We also suggested that papers with little broad interest be published in other journals. The new page charges mostly reflected the increased content in the larger pages.

5. In 1975 we announced that the Supplements would have the same format and style as the main journal, and the charges would be the same as for the Journal. The halftones would be printed on text stock within the papers or on glossy stock at the end of the monthly issues. In general, we would encourage authors of papers longer than 20 pages to accept publication in the Supplements.

6. By 1977 we became concerned with the shelf space occupied by the Journal so the Press found a paper stock that was 17% lighter in weight (with the same see-through), was cheaper, and lowered postage costs. We eliminated blank pages.

7. In 1977 we had to raise page charges by $5 per page but noted that the increase in size of the Journal was paid for by the page charges, not the subscription rates. We announced lower subscription rates for people subscribing to both the Journal and Supplements.

8. In 1978 after three years of constant subscription rates, despite a steady 7-10% annual increase in manufacturing costs, we had to raise the subscription rates. The Letters cost more to produce than the main Journal, so we had to increase its page charges by $10 per page above those for the main Journal.
9. In 1979 we announced a new General Index for years 1971-1975. Dr. Robert Fox at the Press developed a program that has a block of information for each paper giving the title, all the authors, location in the Journal (and microfiche edition) and Supplements, and the subject headings that were attached to the papers. Then these blocks of data are listed for each of the authors in the author index and each of the subject headings. This material is used to generate the annual indexes and the five-year indexes. The Press welcomed corrections. Until digital sources became available, this provided the best way to find past papers.
10. In 1981 we faced the problem that with three issues per month of the main Journal and Letters, the resulting 450-500 pages were too much to bind in a single volume. Henceforth a single volume will consist of two issues, not three. We would continue to publish paper titles in capital letters, but the need to distinguish between e.g. Cs and CS means that we have to allow lowercase letters in titles. Also we would now allow Greek letters, mathematical symbols, italics, and boldface in titles.
11. In 1988 Dr. Dalgarno and I announced a name change from LETTERS TO THE EDITOR to LETTERS. These short urgent publications are not intended for the Editor but for the readers. This change is consistent with usage in other journals and was suggested by the Publications Board.
12. In 1989 I published an Editorial to emphasize that papers on instrumentation were welcome and did not need to be justified, as had been believed, by the astronomical results included. This includes papers on innovative instruments and

techniques, numerical simulations, and analytical solutions. After all, this Journal was founded by George Ellery Hale, one of the greatest pioneers in astronomical instrumentation.

13. Conference papers: in 1990 we faced the four problems with the publication of normal conference papers.

> (1) Normally they are not reviewed, so readers are suspicious of their contents.
>
> (2) The papers are often reviews of material published already and are therefore redundant publication.
>
> (3) Conference volumes are often too expensive for individuals to buy.
>
> (4) Because they are often not parts of series, individual volumes may not be found in many libraries.

Therefore we tried an experiment. For a 1988 workshop on *Impulsive Solar Flares*, organizer Edward L. Chupp and I agreed to send out each of the papers for review and did not tell the referees that each paper was for a conference, so the normal reviewing criteria (importance, accuracy, conciseness, validity, thorough referencing, etc.) applied. Most of the papers were revised and five were not returned. The accepted ones were published in a separate issue of the Supplements and sold for $13. We encouraged others to do this to make conference papers worthy of publication and with a wider distribution.

14. With the growth of international cooperation in astronomy, authors found that they were publishing papers in several different journal, not just one. Each journal had its own style requirements: for instance, some journals use Roman numerals for tables or section headings while others use Arabic numbers. Either are OK, but it is a nuisance to remember all the different journal requirements. Therefore I recommended that the editors of A&A, MNRAS, and myself (for ApJ) convene in Paris (under the auspices of IAU Commission 5) to try to minimize the differences. The Editor of AJ (Paul Hodge) agreed in

advance to abide by the consensus. Surprisingly, we agreed rather quickly on nearly all of the 15 issues. They included simplifications to the reference lists to eliminate all boldface, italics, colons, semicolons, and final periods, saving time and work by the authors and copyeditors. Others included using the international abbreviations for journals outside of astronomy and reasonable acronyms within astronomy, e.g. MNRAS instead of *Mon. Not. Roy. Astron. Soc.* Those changes were published in Abt (1990c). Another change was in reference lists to forego the paper titles and closing page numbers. We realized that those provided useful information, but with 1.5 m references per year in the ApJ alone, those additions would take too much space and time in manuscript preparation.

15. In 1992 we announced that we would accept videos in connection with publication in the Journal. These would be distributed free (by airmail abroad) to all subscribers and in a variety of formats (VHS, PAL, SECAM). Because this became much easier in digital format, I will not go into details here.

16. In 1993 we announced that we would start to distribute to all subscribers CD-ROM disks for large amounts of data. We realized that this format would be replaced by another technology in the near future, in which case the data on CD-ROM disks would then be transferred to the new technology.

17. As our communication entered the age of e-mail and e-mail addresses were becoming standardized, we asked authors in 1993 to add e-mail addresses to their papers.

18. The Centennial Challenge. *The Astrophysical Journal* was founded in 1895 by George Ellery Hale. He saw that the future of astronomy would lie in the application of physics to the study of astronomical objects – hence the word astrophysics. This was done before the discovery of the Bohr atom, the development of quantum mechanics and nuclear physics, the discovery of astronomical gamma- and x-rays, the nature of cosmic rays,

the realization of the extreme size of the universe, and most of the developments of modern-day cosmology. Furthermore Hale had the foresight to realize that astrophysics would need structures that included mountaintop observatories (he founded Mount Wilson and Palomar Observatories). See Section 35 for more about his initiatives. Do we have people now with the vision and foresight that Hale showed? Assuming that all parts of the electromagnetic spectrum are available for study, what will astronomy be like 50 or 75 years into the future? In 1994 I invited essays, which will be reviewed by the Publication Board and become lead articles in future issues of the ApJ. Unfortunately and to my disappointment, no one took up that challenge. Apparently no one in the past century has the foresight shown by George Ellery Hale.

19. As we entered the digital age, we announced in 1996 faster publication for manuscripts submitted in electronic form. Such manuscripts avoid having to be typeset and with copyediting being done by e-mail. This will lead to lower page charges once the process is developed and faster publication. The first copies the authors will see are the typeset page proofs and they can expect not to make changes at that stage.

20. At that time we called attention to the thinner issues, not caused by fewer papers but by paper stock that is 22% thinner, while retaining 93% opacity. The lighter weight also caused smaller postage costs. When we changed typesetters, we went to a process of reducing the font size by 5% to match the previous font size. Now to save that expense, we are using the larger font size, making the text marginally easier to read. To avoid increasing the expense we have opted for smaller margins and adding one line per page.

Finally in 1996 we decided to terminate the microfiche series (we had only 69 subscribers for that) because we planned to start the on-line Journal in January 1997.

21. The ApJ has always had one or more Associate Editors to advise the Managing Editor, or now, the Editor-in-Chief. In 1996 I realized that I could no longer select all of the referees and oversee the reviews for all the papers. Therefore I converted the six Associate Editors into Scientific Editors. They would select the referees and oversee the reviewing for papers in their fields. In 1996 I added six more with the approval of the AAS Council.

    They were selected to be experts in different fields of astronomy, e.g. solar system, stars, gaseous nebulae and interstellar matter, galaxies, cosmology, and high-energy objects. For each of incoming manuscript, I simply decided to which Scientific Editor it should be sent. Later some of those were abroad because many manuscripts came from abroad.

22. Prior to 1967 the halftones were printed on glossy paper within the papers, but that involved the expense of cutting each 32-page signature by hand and inserting the halftones. Later all halftones were printed on glossy stock at the ends of each issues. With improved paper stock, some halftones could be printed on text stock within the papers. In 1998 the Press found a semi-gloss paper that allowed high-resolution halftones within the papers. The new paper had the same opacity (92.5%) as the recent text stock. It had 95% of the weight, reducing postage costs. It had 71% of the thickness so the shelf-space will be less. It was first tried in the January 1998 AJ and then used in the 10 June 1998 issue of the ApJ. The Letters continue to have a 4-page limit, which would then include halftones.

We were indebted to the staff of the University of Chicago Press, particularly its Director Robert Sherrill, for suggesting some of those improvements and implementing all of them.

Meanwhile, the possibility of online publication occurred and the AAS Executive Director, Dr. Peter B. Boyce, took the imitative to obtain a grant for $500,000 from the NSF to create that version. It appeared in 1996 as one of the first scientific publications to be available on-line. By 2015 the printed version stopped. A&A and MNRAS did so at about the same time.

It is evident that no ApJ Editor before or after made so many (by a large factor) improvements and changes. I would have been willing to continue as Editor-in-Chief beyond 1999, but I gather that one member of the Publications Board wanted to replace me and campaigned for that. Nevertheless the Publications Board selected another person. The first candidate was a disaster. He thought that he could delegate most of the work and only spend part time on the Journal. The next candidate was Robert C. Kennicutt, who took over in 2000. However he soon moved to Cambridge, England and operated from there. He thought that he could do with fewer than 15 Scientific Editors but soon realized that he needed that number. After six years he resigned and Ethan Vishniac took over in 2006.

I was helped in Tucson most of the time by 3-5 assistants at a time, of which Ruth Perry, Iona Hoehn, and Janice Sexton were the most loyal and helpful. In particular, Janice helped in our gradual transition to computer handling of manuscripts. It was a pleasure to work with them. We answered all letters, including from cranks. To cranks we replied that in scientific papers all statements had to be substantiated. I wrote the bodies of all the letters (except for the acceptance form letter) and my assistants added the opening, closing, and wrote my name ("Helmut"). I felt that when a referee spent many hours reviewing a paper, the editor should spend at least 5-10 min. writing a personal letter of thanks.

Because the budget of the Journal was roughly ten times that of the AAS, I had to attend all Council meetings for 29 years, i.e. 58 meetings. I think that I gave a research paper at each meeting. Together with going to ~25 scientific meetings abroad, 14 trips to China to help them get started in astrophysics, ~30 trips for pleasure, seven observing runs to Chile, and roughly two observing runs every month, I did a lot of traveling.

An overview of changes, growth, and records set was published in Abt (1995b). It shows the major changes since 1895. The Journal annual content was flat for the first 35+ years but then rose rapidly since the 1930s. That corresponded with the growth of atomic and nuclear physics. There was a gap during WWII, but a constant logarithmic increase of 8.8% per year with no additional increases during the space age.

Among the records listed in that review was one person who was an author on 137 papers in five years or one paper every two weeks – mostly written by his many coauthors and often with redundant material in different journals. One of the most prolific authors was E. C. Pickering who wrote dozens of trivial 10-15 line papers. The outstanding authors of substantial papers were Otto Struve and S. Chandrasekhar. The paper that took the longest to get published took 11 years after submission; the referees took only two months during that interval. The shortest time (before my time) had the same dates of submission and acceptance; it might have been correct if the author and editor took more time. The longest papers were lists of radio sources (503 pages), a catalog of QSOs (496 pages), and the identification of solar chromospheric lines.

I had developed a different approach to editing than that of other editors. It is explained in an essay not yet published called *Principles of Editing*. One of the main points is that authors have rights also and editors should not act as dictators. Another main point was that the referee was always told how the manuscript was changed in response to the referee's report and how it was handled thereafter. The only area where I differed with Chandra was in the selection of referees. He conscientiously selected an opponent of the author of a controversial manuscript because he was more concerned with the accuracy of published papers than with loosing good papers. I found that if an opponent was selected, the reviewing never finished or the paper was never published.

I developed a method in which if an author openly contested another author, I would ask the contested author for his/her comments but not to act as a referee, i.e. not to make judgements regarding publication. Then those comments were sent to the new author for a written reply. The two sets of comments were then sent to a neutral referee for the referee's use in evaluating the manuscript. Both the contested author and referee were kept informed about the progress toward publication.

I remember one author who was rightfully upset (before my editorship) when a paper appeared in print that contested his work and he had no prior knowledge of it. Such a procedure wastes time, journal space, the patience of authors, and leaves the readers confused. The Editor must always scan manuscripts enough to detect criticisms of others, either by name or subtly by content. Then the Editor should apply the above procedure for contesting papers.

Surprisingly, I do not recall ever rejecting a paper. I believe that the reviewing process should continue until the author realizes that there were problems with the paper or the referee is convinced that the paper is correct. Also, I always informed the referee how the manuscript was changed in response to the referee report and whether the paper was accepted for publication.

I did not think that a referee should be asked for more than two reviews of a manuscript. If the referee is asked for more reviews, the referee is likely to get tired of it and perhaps allow publication when he/she had lingering doubts. If an agreement was not reached after two reviews, I consulted a second referee, who received anonymous copies of the previous reports and the author's replies. In other words, the second referee was acting as an arbitrator. In very rare cases a third referee was consulted. But science in a controversial area can be difficult. I always admired a referee who said in effect "I do not believe the author's results but I cannot prove them wrong so I think the author should have the right to publish his/her conclusions".

In an American court, the judge would never decide a case after hearing only the prosecutor's statement. The judge would insist on hearing both sides and letting the defendant and prosecutor question each other. In the same way, an Editor should never decide for or against publication of a paper by a qualified author after reading only the referee's report. Several times I experienced the opposite to my own papers.

In recent years I always identified myself as a referee, but I do not recommend that all referees do that because some authors can be vindictive. If the referee needs to submit a negative report and then later needs to seek a grant, promotion, or new position, the rejected author may seek revenge.

23

MULTIPLICITY OF SOLAR-TYPE STARS (1976-2006)

I wondered what fraction of stars like the Sun have stellar companions. Double, or binary, stars can be discovered in several ways. One is to obtain high-dispersion spectra of stars and look for variations in their Doppler or radial or line-of-sight velocities; those are called spectroscopic binaries. They mostly have periods of less than a day to 100 years. Another is to look for two stars close together and measure their separations and orientations. Those are called visual binaries, and generally have periods of 10 to 1000 years. A third is to make accurate position measurements of single stars and note that they move in orbits around the centers-of-gravity with unseen companions. Those are called astrometric binaries. During the first half of the 20th century, the major observatories (Lick, Mt. Wilson, Yerkes, Dominion Astrophysical Obs., and the Dominion Obs.) devoted most of their observing time to discovering and measuring spectroscopic and visual binaries. They found many stars to be double.

We know that the Sun does not have a companion star but has eight or more planets and other objects. So I used the Kitt Peak 2.1m coude spectrograph to obtain at least 20 spectra for each of 135 solar-type (F3-G2 IV or V) stars brighter than V = 5.5 mag. The results were published in Abt & Levy (1976). We made heavy use of older radial-velocity measures published elsewhere to augment our measurements. Orbital elements were obtained or collected for 21 spectroscopic binaries, 23 visual binaries, and 25 astrometric pairs. The mean accuracy of the radial-velocity measures was ±0.6 km s^{-1}. The final result was that the frequencies of single: double; triple; quadruple systems was 42:46:9:2 %. That meant that more than half of the nearby stars were multiple stars, a very surprising result. Some of this research was funded by the Research Corporation when KPNO could no longer afford to give free observing time to me but had to sell it.

This result was challenged by Branch (1976), who pointed out that there was a bias for doubles because some of those with total magnitudes brighter than V = 5.5 mag. had primaries fainter than that magnitude. Correcting for that by deleting 20 binary stars, the frequencies of singles:double:triples:quadruples is 45:46:8:1 % (Abt 1978). That still means that more than half were multiple.

The Swiss astronomers headed by Michel Mayor (Duquennoy & Mayor 1991) repeated that experiment but with improved equipment having an accuracy of ±0.3 km s^{-1}. They selected a sample of F7-G0 IV-V, V, and VI stars within 22 pc (parsecs) according to the parallaxes of Gliese (1969). They found that at least two-thirds of the solar-type stars are multiple.

Unfortunately their use of Gliese parallaxes was a mistake. When the Copernicus measures became available, more than half of their sample fell outside 22 pc and corrections had to be made (Halbwacks et al. 2004).

The first claimed exoplanet found was probably γ Cep (HR 8974) by Campbell et al. (1988), who made measures with an accuracy of ±0.013 km s^{-1}. Perhaps a more convincing case was 51 Peg (HR 8729) by Mayor & Queloz (1995). For that discovery Mayor received the Noble Prize for Physics in 2019. To date more than 4000 exoplanets have been discovered. In a sense, this all started with the Abt & Levy (1976) paper.

24

EXOPLANETS

There remains a problem with which I have been involved, namely the origin of exoplanets. Most observers assume that the discovered exoplanets were formed in disk systems like the solar system. The idea is enhanced with discoveries made with the Kepler spacecraft, which discovers exoplanets by detecting transits of their primary stars. It favors co-planar exoplanets. But there is another way to form exoplanets, which is as self-gravitating objects such as stars. Boss (2003) and others have shown that objects with masses as low as those of planets can form in that way. Then in clusters the individual objects become captured as doubles. The observational difference between independent self-contracting objects and disk objects is that the former are spherically distributed, while the latter are in disks. Abt (2010) showed that the period distributions for the two models do not distinguish between the two, and suggested that most exoplanets were formed like stars are formed, not in disk systems. Tremaine & Dong (2012) analyzed the data to that date and concluded that many exoplanets were formed in spherical system.

An observational limitation is that the exoplanet observers have a self-imposed limit, namely that if the velocity variations of a primary star exceeds 2 km s^{-1}, they stop observing it because that star will not have planets, but stellar or black dwarf companions. Therefore the frequencies of doubles among stars, black dwarfs, and exoplanet systems remain unknown.

A large sample of exoplanets (Abt 2010, 2011) showed that more than half have eccentricities > 0.1, unlike the solar system in which only one planet has e > 0.1. The eccentricities and semi-major distances of stellar, brown-dwarf and exoplanet systems are the same, showing that most of the exoplanets were formed individually as self-gravitating objects and then become associated in capture systems. There are a few known stars with multiple planets measured in three dimensions that show them to be spherical. Furthermore, becoming associated in captures explains why exoplanets are rare among metal-poor stars because like in stellar companions around metal-poor, the peak period is at 900 days. These statistics from 2011 show that for exoplanets discovered by that date, most were formed individually and by captures but not in disk systems. My attempt to further publish these ideas was met with rejections by referees and editors without my being allowed to respond to the referee reports.

25

CATALOGS

Another thing that I did was to do things that were needed for astronomy that other did not do. One, mentioned above was to compile (with Eleanor Biggs) three General Indexes for the ApJ. Another, once I had the process in place, was to make a General Index for the AJ for 1944-1975.

Then I realized that it was difficult to find the sources of all the published radial velocities of any given star. Therefore I scanned all the astronomical journals and observatory publications to find papers that gave new measurements of radial velocities. I indicated which tables for Eleanor to copy and she typed onto IBM cards the velocities, number of measures, references, etc. We included the star's identification name, HR and HD numbers, positions, and spectral types. That material was sorted by position and transferred to a magnetic tape so that only the magnetic tape needed to be sent to the publisher. That came out as the *Bibliography of Stellar Radial Velocities* by Helmut A. Abt and Eleanor S. Biggs (1972). It was probably one of the first published book based entirely on a digital text. I still have printed copies available free to anyone. After the establishment of the Astronomical Data Center in Strasburg, France, they took over the inclusion and extension of those data to date.

Another large compilation was the unpublished Mt. Wilson radial velocities. Some of their astronomers, such as Ralph E. Wilson, were only interested in the mean velocities of stars because they were interested in the galactic motions of stars. Therefore they obtained five of more radial velocities of each of about 15,000 stars (Wilson 1953), enough so they felt comfortable that they had obtained the mean velocity of each star. However, for those of us interested in spectroscopic binaries, we wanted to know each individual measurement, not just the means. The individual velocities were on card files in the attic of the Mt. Wilson Observatory building at 813 Santa Barbara St., along with the times of observation in Pacific Standard Time. To aid the astronomers to use standard time, the whole mountaintop stayed on Standard Time when Pasadena and the rest of California went on Daylight Time.

In the 1960s I went to Pasadena monthly to collect the individual radial velocities. At the same time there, I copied all of the Mt. Wilson spectroscopic log books; in case of emergency to the Mt. Wilson Observatory buildings, I have those duplicate logs. I converted the dates to UT times. In the published catalogs (Abt 1970, 1973) I listed the star names, HD, ADS and Wilson numbers, the 1900 coordinates taken from the HD catalog, the measured velocities, the probable errors if the plates were measured more than once, and the spectroscopic dispersion used. Those catalogs have been heavily cited by observers of spectroscopic binaries.

In another classification project to provide useful information for others to use, I classified 584 stars previously lacking MK types and listed in *A Supplement to the Bright Star Catalogue* (Hoffleit et al. 1983). The Hoffleit et al. catalog was compiled to add stars brighter than V = 7.1 that had not been included in the *The Bright Star Catalogue* (Hoffleit & Jaschek 1982).

26

PUBLICATION STUDIES

In 1980 we had a KPNO Director who was having difficulty in balancing his budget, so he proposed closing all the small telescopes because, he said, "they are used only by students and did not produce important research". I questioned that, so I did a study. I collected data on the 445 papers published in a previous five-year interval by staff and visiting astronomers for all the Kitt Peak stellar telescopes and the 4179 citations to them in five years, compiled by the Institute of Scientific Information in the *Science Citation Index.* What was needed in addition was the costs of building those six telescopes and the instruments on them, and the operating cost for those telescopes. Buddy Powell and C. E. Johnson of our observatory staff provided those data. Included in the results was useful information, such as the initial costs of telescopes varied as the 2.4 power of the aperture. All that detailed information was published in Abt (1980). I included the annual costs of operating the telescopes plus 1/75 of the initial costs, assuming that telescopes had useful average lifetimes of 75 years.

The result was that the number of citations per dollar for the five small telescopes was several times larger than for the 2.1m telescope. That means that the science per dollar is greater for small telescopes than for larger ones, although there is science that can only be done with a larger telescope. Those results stopped all attempts to close the small telescopes. It also showed the power of publication studies.

I realized that the use of lists of published papers and citations can produce important information about astronomical research and how astronomy progresses. In the following 40 years I published approximately 75 papers on publication studies. I will summarize below 46 of those.

1. "Some Trends in American Astronomical Papers" (Abt 1981b). The numbers of American papers published was constant in 1910-1940. After WWII they doubled every 7.8 years. Papers became longer, but the papers per astronomer decreased by 40%.
2. "Statistical Publication Histories of American Astronomers" (Abt 1982). American astronomers published an average of five 1000-word research papers per year but a few published 15-20 research papers per year.
3. "At What Ages Did Outstanding American Astronomers Publish Their Most-cited Papers?" (Abt 1983a). For 22 outstanding astronomers whose careers terminated before 1970, their total citations per year formed a broad fairly-flat distribution from ages 30-80, with a gradual peak in their 60s.
4. "Citations to Single and Multi-authored Papers" (Abt 1984). On the average, the number of citations to a paper increased with the number of authors, but not proportionally. Doubling the number of authors produced a 50% increase in citations. Longer papers averaged more citations than shorter ones, but not proportionally. Double-length papers increased the number of citations by 50%.
5. "Are Papers by Well-known Astronomers Accepted for Publication More Readily Than Other Papers?" (Abt 1987a). Papers by well-known astronomers take slightly longer to be accepted, but their acceptance rate is 95% compared to 83% for others.
6. "Reference Frequencies in Astronomy and Related Sciences" (Abt 1987b). The number of references in papers are directly

proportional to their lengths. Astronomy papers have more references than papers in physics, chemistry, and geophysics because they are longer.

7. "What Happens to Rejected Astronomical Papers?" (Abt 1988a). Of manuscripts submitted to the AJ, ApJ, and PASP, eventually 90% are accepted by the same journal. Of the remaining 10%, two-thirds are never published and one-third are published in other journals. Outside of the physical sciences the acceptance rates are 10-30%. That is because journals that depend upon subscriptions for most of their income have to decide before each year starts how many pages they can publish and generally set that number low.

8. "Growth Rates in Various Fields of Astronomy" (Abt 1988b). During 1970-1985 worldwide, the numbers of astronomical papers doubled every 18.3 years. The rate depended inversely upon distance of the objects: they decreased in planetary science during that time but doubled every eight years in galaxies and cosmology. During that interval the fraction of papers from American authors decreased from 38% to 32%.

9. "Trends Toward Internationalization in Astronomical Literature" (Abt 1990a). During 1970-1990 in two American journals and three non-American journals, the fraction of papers from more than one country increased by 30%.

10. "Publication Characteristics of Members of the American Astronomical Society" (Abt 1990b). For the 4995 members of the AAS in 1989, Full members publish 3.34 papers per year, or 0.5 normalized (citations divided by the number of authors) papers.

11. "Science, Citations, and Funding" (Abt 1991). I responded to the claim published in *Science* magazine that 55% of papers are never cited. The actual numbers for research papers is 5.1±1.0% in astronomy and 8.1±1.2% in physics. The

published 55% fraction included abstracts, book reviews, announcements, errata, editorials, and obituaries.
12. "What Fraction of Literature References are Incorrect?" (Abt 1992a). For 1009 recent references in ApJ, 12.2% had errors; 0.4% could not be found at all; 3.0% were found by using the indices; 8.3% had errors in names, journals, or locations such that SCI (Science Citation Index) missed them. However, SCI matches citations with publications so SCI missed only 3.6% of the references.
13. "Publication Practices in Various Sciences" (Abt & Garfield 1992). A study of papers published in 1990 in eight sciences showed that the fraction of multi-national authorships ranged from 2% in psychiatry to 26% in astronomy and physics. The average paper lengths ranged from 4.6 pages in medical papers to 13 pages in some fields. Revision rates ranged from 8% in math to 100% in geophysics, and times of publication ranged from 200 days in physics to 600 days in math.
14. "The Growth of Multi-wavelength Astrophysics" (Abt 1993). During 1962-1992 the papers that consisted of new observations numbered 57%, re-discussions of data 12%, theoretical and laboratory 28%, and instrumentation 4%. Optical papers changed from 79% to 46%, gamma rays became 1%, x-rays became 9%, UV 6%, IR 15%, and radio remained at 20%.
15. "Changing Sources of Published Information" (Abt 1995a). During 1972-1992 the fraction of research papers published in journals stayed steady at 78%, those in observatory publications decreased from 12 to 1%, in conference proceedings from 1 to 10%, and in monographs (books) from 6 to 4%.
16. "What Fraction of Astronomers Become Relatively Inactive in Research after Receiving Tenure?" (Abt & Zhou 1996). After receiving tenure, 1.8% published no further research papers, 19% published up to half of their previous rate, 33% published

half to one times their previous rate, and 48% published more than their previous rate.
17. "How Long are Astronomical Papers Remembered?" (Abt 1996). For papers published in 1954, citation counts show a half-life of 29 years. The half-lives are 35% for observational papers and 22% for theoretical ones.
18. "Why Some Papers Have Long Citation Lifetimes" (Abt 1998). The main reason why astronomical papers have a half-life of 29.3 years and for physics it is 10.6 years is because astronomy has been growing faster, meaning that there are more researchers now to refer to the older papers. After allowing for different growth rates, papers in astronomy, chemistry, geophysics, physics, and general science all have a mean half-life of about 8 years.
19. "American Astronomical Society Centennial Issue" (Abt, ed. 1999a). I asked 50 outstanding astronomers, mostly American, to select the most important papers in their fields that were published in the past 100 years in the ApJ or AJ. and then to explain why. The resulting 53 papers were published in a special issue of the ApJ in their original formats and were followed by the commentaries.
20. "Do Important Papers Produce High Citation Counts?" (Abt 1999b). For the 53 papers selected to be among the most important in the past century, we found that those published before 1950 had 11 times the citation counts as control papers and those after 1950 had 5.1 times the citation counts of control papers. Thus important papers produce higher citation counts than average papers.
21. "Millennium Essay: Astronomy Publication in the Near Future" (Abt 2000a). During 1970-2000 the number of American astronomical papers increased linearly with the number of members of the AAS and showed no additional increases due to the completion of new telescopes, the launching of new spacecraft,

new technical or scientific capabilities, equipment sensitivities (e.g the advent of CCDs), or breakthroughs in computational or publishing techniques. When new possibilities occur, astronomers may publish better or more important papers, but not more. Thus if an organization wishes to increase its publication numbers, it should hire more astronomers, not acquire more equipment.

22. "The Most Frequently-cited Astronomical Papers Published During the Past Decade" (Abt 2000b). For the 100 most-cited papers during 1988-1997, 49% were extragalactic, 10% were galactic or interstellar, 36% were stellar, 1% were solar, and 3% were solar system. Of the 100 papers, 47% were observational, 49% were theoretical, and 4% were instrumental.

23. "The Productivity of Ground-based Optical Telescopes" (Abt 2001). By scanning 1996 and 2001 papers in five European and American telescopes, we found that 82% of the papers and 75% of the citations came from telescopes 4m in aperture. The mean citations per paper is less than the ratio of telescope apertures.

24. "Is the Relationship between Numbers of References and Paper Lengths the Same for All Sciences?" (Abt & Garfield 2002). In each of 41 journals in the physical, life, and social sciences, the relations between number of references and normalized paper lengths are the same for all journals in each science to ±17%. This allows us to inter-compare citation counts with that accuracy.

25. "The Productivity and Distribution Times for Conference Proceedings" (Abt 2002). For 300 astronomical conferences, I counted the times between the last days of the conferences and receipt of the printed proceedings. The mean times were 10 months of the ASP Conference Series, 11 month for IAU Symposia, and 14 months for others. A total of 29% were not published in 1.5 years.

26. "What Factors Determine Astronomical Productivity? (Abt 2003). The number of international papers per IAU member has remained constant during the past 30 years at 0.9 ± 0.08 papers per year. It shows that the productivity depends only upon the number of astronomers and not on the availability of equipment, spacecraft, detector speeds, or increasing computer power.
27. "The Scientific Output of the International Ultraviolet Explorer During Its Lifetime" (Abt & Boonyarak 2003). We counted the citations to the 3435 papers produced with IAU in 1978-1996. They show a 61% increase per year during 1980-1994, due partly to including extragalactic research. IAU showed an unusually high productivity even when NASA decided to turn it off in 1996. During the last six years it averaged 16.3 ± 2.5 citations per paper per year.
28. "What Kinds of Astronomical Paper Are Still Referenced 50 Years after Publication?" (Abt & Boonyarak 2004). We considered the 603 papers published in 1950-1951 in American and European astronomical journals. During 2000-2003, 20.4% of the papers still show citations. The 50 papers with 3-39 citations are listed. They show a citation peak 36 years after publication ("late bloomers") compared to a peak after 5-6 years for other papers.
29. "Estimated Completeness of the Science Citation Index" (Abt 2005a). I looked at the six 1997-2004 journals that publish the bulk of the solar physics papers and found that for five papers published in 1997 *Solar Physics*, 99% of their citations are listed in SCI. Therefore the low Impact Factor for that journal is not due to citations missed by SCI.
30. "A Comparison of Citation Counts in the SCI and the NASA Astronomical Data Systems" (Abt 2006). The SCI and ADS agree for 85% for a large sample of references. SCI includes

more papers in physics and chemistry, while ADS includes some conference papers. Each misses 1% of the citations.

31. "National Astronomical Productivities" (Abt 2005b). I looked at the authorships of 6902 astronomical papers published in 2003 in the 13 astronomical journals with the highest Impact Factors. I found that for 23 of the 30 developed countries and seven of the developing countries that the numbers of papers is roughly (±34%) proportional to their Gross Domestic Products (GDP).

32. "The Frequencies of Multinational Papers in Various Sciences" (Abt 2007a). The multinational authorships from the top four journals in each of 16 fields of science were studied. The frequencies ranged from 13% in surgery to 55% in astronomy. The most likely explanation is that sciences that study internationally-distributed objects (astronomy, geophysics, biology) have more multinational papers than those confined to single laboratories or hospitals.

33. "The Publication Rates of Scientific Papers Depend Only on the Numbers of Scientists" (Abt 2007b). In astronomy, chemistry, geophysics, math, and physics during the past 30-35 years the numbers of papers are proportional to the numbers of members in their scientific societies with no jumps due to improved facilities or instrumental breakthroughs. The papers may become better, but there are not more of them.

34. "The Future of Single-author Papers" (Abt 2007c). For the sciences astronomy, biology, chemistry, and physics during 1975-2005, the frequencies of single-author papers fit decreasing exponential curves but never reach zero.

35. "Reference Sources in Research Literature" (Abt 2009d). I looked at a large sample of references in A&A and ApJ during approximately decade intervals during 1952-2009 and found the following: references to journals increased from 76 to 90%, monographs decreased from 7 to 4%, conference proceedings increased from 1 to 7 to 3% with a peak in 1980, in-house

publications decreased from 12 to 1%, preprints increased slightly from 1 to 2%, reviews increased from1 to 2%, private communications decreased from 2 to 0%, and theses increased slightly from 0.5 to 1%.

36. "Reviewing and Revision Times for the Astrophysical Journal" (Abt 2009c). For 251 manuscripts submitted to the ApJ in 2006, 6% were rejected, 5% were withdrawn, and 88% were accepted for publication. Of the accepted manuscripts, 30% were reviewed once, 58% were reviewed twice, and 12% were reviewed 3-5 times. The first review averaged 31 days and the second 44 days. Important papers (with 31-193 citations in two years) were not reviewed nor revised more promptly than less cited papers. Only in the field of high-energy objects were the papers revised marginally more promptly than others.

37. "Astronomical Publication Rates in the US, UK and Europe" (Abt 2011). I explored the growth rates of astronomical research papers in the US, UK, and four European countries (France, Germany, Italy, and the Netherlands). I counted pages in four major journals and corrected for format differences and contributions from other countries. The UK lags behind the US by 10 ±1 years and the European countries lag behind the US by 20 ±1 years.

38. "Scientific Efficiency of Ground-based Telescopes" (Abt 2012). I scanned six major journals in 2008 for all the 1589 papers based on ground-based optical/IR telescopes worldwide. I then collected citations to them in three years, the most-cited papers, and the annual operating costs. They were grouped into four groups by telescope apertures. Those with apertures > 7m averaged 1.29 more citations than those from 2-4m telescopes, which is a small return for a factor of four in operating costs. Among the 17 papers with >100 citations in three years, only half came from the large-aperture (>7m) telescopes. We wondered why they did so poorly and suggested reasons why.

Papers based on archival data (e.g. the Sloan Digital Sky Survey) produced 10.6% more papers and 20.6% more citations than those based on new data. Also, the 577 papers based on radio data produced 36.3% more papers and 33.6% more citations than the 1589 papers based on optical/IR telescopes.

39. "The Research Use of Astronomical Monographs" (Abt 2014). I studied the 135 monographs (books) listed in the 2000-2003 issues of *Physics Today*, excluding conference proceedings and textbooks, and counted citations to them in 2000-2013. I found that 67% of them received <2 citations per year, compared with 41% for ApJ papers. The average citation counts were like those for ApJ papers. ADS also counts downloads in 14 years. The numbers are 181 ±27 for monographs and 633 ±47 for ApJ papers. The mean total citation count in 14 years range from 0 to 711 citations. These numbers do not depend on the number of book reviews or the scientific stature of the authors. I am unable to predict whether a monograph will be successful or not. The reasons for these statistics are (1) most monographs are judged to be too expensive for individuals or libraries to buy, (2) many readers prefer concise reviews, such as in online searches, and (3) most monographs are not available online. However, monographs are useful to learn a field.

40. "The Lifetimes of Astronomers" (Abt 2015). For members of the AAS I collected data from (1) 489 obituaries published in 1981-2015, (2) 127 deceased AAS members but without published obituaries, and (3) a sample of AAS members without published obituaries or not known to the AAS as being deceased. The most frequent lifetime was 85 years. Of 674 members 11.0 ±1.3% lived ≥90 years. For the general population the peak lifetime was 77 years, showing that astronomers live an average of eight years longer, perhaps because astronomy is not as dangerous or stressful an occupation as some others are.

41. "At What Ages Did Astronomers Write Their Most Important Papers?" (Abt 2016). In 1983 I found (Abt 1983) that the most productive ages for research astronomers was 40-75 years, contradicting the frequent statement that a scientist's best work was done before the age of 35 years. Now most astronomers work in teams, unlike the individual research usually done previously. How has this affected the productive careers of astronomers? A study of 14 recent Russell Lecturers showed a peak in productivity at age 33 years and more than half that productivity during 25-56 years, in agreement with the results from other scientists. Nevertheless, 33% of their best work was done after 50 years of age.
42. "Citations and Team Sizes" (Abt 2017a). I explored whether small or large teams produce more citations by looking at 1343 papers in the Jan.-Feb. 2012 issues of A&A, ApJ, and MNRAS and then counting citations in 4.5 years. Large teams produce more citations by a factor of 2, but is that due to self-citations? I found that self-citations had a linear range from 12.7% for the smallest teams to 45.9% for the largest. However that is not enough to explain the factor of 2. But if we look at normalized citations (citations per author), small teams do better than large ones by a factor of 6. Large teams tend to produce more data but small ones emphasize physical processes.
43. "The Most Productive Years of Average Astronomers" (Abt 2017b). We learned previously that geniuses and outstanding astronomers have productive peaks in their 30s but continue to be productive late in their lives. This time we consider average astronomers, namely 25 AAS members who died recently. Their productivities peaked in their 40s and published half of their citations peaks in 28-67.
44. "What Fraction of Papers in Astronomy and Physics Are Not Cited in 40 years?" (Abt 2018). For research papers it is 1.4% for astronomy and 1.5% for physics. The peak ages of productivity

are 40.3 and 36.6 years, respectively. Also, 43.0 and 33.7%, respectively, are published after the age of 50 years. Therefore physicists peak four years earlier and published 9% fewer papers after age 50 years.

45. "The Lifetimes of Astronomical Papers and the Completeness of the ADS" (Abt 2019). Based on citations, we determined how long astronomical papers are remembered. Papers from 1955 had a half-life of 71 years, for those from 1960 it was 25 years, and after 1970 it was a steady 10 years. This tells us that astronomical results are now quickly appreciated and quickly replaced. However, astronomical papers are receiving increasing numbers of total citations. Tests for astronomical journals show that the ADS has a 94% completeness, although that does not include most observatory publications and books.

46. "Publication Changes during the IAU History" (Abt 2019). Early in the 100-year history of the International Astronomical Union, one had to be able to read the astronomical papers in eight languages (Chinese, English, French, German, Italian, Japanese, Russian, and Spanish) but now 99% of the 160 astronomical journals in the world are printed in English. The world's astronomical publication has increased exponentially. Observatory publications, conference proceedings, theses, and private communications have nearly disappeared in research usage in favor of journals.

27

CHINA

I find China to be the most fascinating country in the world, partly because of its ancient building, beautiful areas, artistic objects, and early discoveries and accomplishments that predate those of the west by 2000 years. Here are my favorites:

- The Qin Shi Huang terracotta army of 209 BC discovered in 1974 east of Xi'an with ~8000 life-sized soldiers, 130 chariots, and 670 horses, the most impressive scene I have ever seen (Fig. 27-1).
- The beautiful Huangshan (Yellow Mountain), a craggy mountain with pines trees growing in impossible cracks, the favorite subject of Chinese artists.
- The Stone Forest east of Kunming, consisting of thousands of limestone formations covering 500 km^2; a UNESCO World Heritage Site (Fig. 27-2).
- The records of hundreds of major discoveries in science, engineering, and agriculture made approximately 2000 years before the west; these were published in the 58 volumes by Joseph Needham and colleagues. See Section 28.
- The stark beauty of the Tibetan Plateau and the realization that China displaced the Buddhist repression of the people to 15th

century conditions and slavery to one of free education, modern medical help, and a contemporary poverty-free economy.
- The Li River in southern China, centered at Guilin and having 200m karst peaks.

Fig. 27-1. Pit 1 of the Qin Si Huang's 206 BC terra cotta army

Fig. 27-2. The Stone Forest near Kunming, China. It consists of thousands of limestone pinnacles covering 500 km².

- The Han Yangling Mausoleum of 153 BC, north of Xi'an, with 81 burial pits (20 have been opened) discovered in the 1990s and containing thousands of miniature-sized (1 foot = 0.3 m high) pottery soldiers and tens of thousands of animals.

There are interesting places that I have not seen, such as

- Rainbow Mountain (Zhangye Danxia) in Gansu Province, probably the most colorful geological formation in the world (Fig 27-3).
- Zhangjiajie National Forest Park consisting of thousands of sandstone columns, up to 200m high, and caves, in southern China. Named a UNESCO World Heritage Site in 1992.

- Dun Huang (Mogao) Caves or the Thousand Buddha Grottos are a group of 500 caves carved into the rock in the 4-14th centuries and containing 50,000 Buddhist documents, 2,000 clay sculptures, and the Diamond Sutra, the oldest completely-printed book (dated 11 May 868 AD), written in Sanskrit. It is in northern China and is also a UNESCO World Heritage Site.

Fig. 27-3. Rainbow Mountain (ZhangYe National Geopark) in Gansu, China. A road passes through it.

The Chinese Cultural Revolution lasted from 1966 to 1976. It was Mao Zedong's attempt to reverse some of the severe problems that China had, such as a rapidly-growing population, insufficient food, and too great a difference in incomes. He made many of the scientists, professors, artists, and technical people spend years on farms, so they fell badly behind the rest of the world in science, industry, technology, etc.

Part of the Chinese attempt to catch up in astronomy was an invitation by the Academia Sinica in 1985 to hold a meeting on close binary systems in Beijing in November 1985. Kam-Ching Leung (Univ. Nebraska) convinced the NSF to fund the expenses of 10 American astronomers to go to the major centers of research and take part in the meeting in Beijing. The participants were: Helmut Abt, Anne Cowley, Bob Koch, Yoji Kondo, Kam-Ching Leung, Doug Lin, Ron Taam, Ron Webbinck, and Bob Wilson; Bohdan Paczynski was invited but did not go. Doug Lin was a huge help to us because he had spent his youth in China before moving to Canada and the US. He could tell us much about the significance and history of what we saw. The stops were Hong Kong (1 day), Guangzhou (1 day), Kunming (3 days), Beijing (8 days), and Nanjing (3 days).

The flight from San Francisco to Tokyo took 11 hr, 15 min.; the plane was one-third filled. The flight from Tokyo to Hong Kong (Excelsior Hotel) took five hours. There were no customs checks.

In Hong Kong we did the usual tourist things: took the tram up to Victoria Peak in the evening to see the lights of Hong Kong. We took the Star Ferry (10¢) to Kowloon – the mainland part of Hong Kong for shopping - and had Peking Duck for dinner (my favorite food). The Hong Kong airport at that time had a short runway in Kowloon where the planes flew to the end of the runway and then many meters over water before rising. Scary!

We flew (CAAC 304, Boeing 727) on 3 Nov. to Guangzhou, China (formerly called Canton), the port where the Pearl River meets the Pacific Ocean. It is now a chief commercial city along with Shanghai. The local language (and food) is Cantonese, although all the young people now can speak the official language of Mandarin. Another local language up the coast and in Taiwan is Fujian. In Guangzhou we visited the Zhenhai Tower (built in 1380) with its exhibit of artifacts from a 400 BC tomb. We also visited the Guangzhou Friendship House and I drooled over the jade carvings.

In Kunming we visited their observatory that has a 1m optical telescope with a coude, a satellite tracking telescope, and a radio telescope for observing solar flares. The Observatory Director was Tan Huisong, who had observed at McDonald Obs. Later he worked with me at Kitt Peak for a couple years, and retired in San Francisco where his children lived. Our 12-course dinner included quail eggs and fungus, roast pigeon, rose petal soup, etc. The spectacular place near Kunming is the Stone Forest, 90 km east and that consists of a 500 km^2 area of eroded limestone columns (Fig. 27-2). Kunming has many minority groups with their own languages, foods, music, musical instruments, dress, Buddhist temples, and handicrafts. Among the latter are imaginative batiks. I bought silk cloth for $4 per meter.

In Beijing we stayed at the Friendship Hotel, built by the Russians and looking like a conservative Russian building. The meeting was held in the hotel. We were on the 7 and 10 pm TV news because this was one of the first international scientific meetings after the Cultural Revolution. We saw parts of the meeting on TV. In the evening at the Friendship Store I bought a serpentine carving of birds, flowers, and trees for $220. It was a wedding gift from southern China (because of the birds portrayed). Later in Tucson it was appraised at $4000.

The next cold windy day (25° F) we visited the Great Wall at Badaling and then the Ming Tombs. Of the 16 Ming Dynasty (1368-1644) Emperors, 13 were buried there. Only one has been opened: Tomb Ding Ling, of Emperor Ding Ling who ruled during 1573-1620. But as soon as the outside air came in, the fabrics and other fragile materials disintegrated. Therefore the Chinese have refrained from opening other tombs until they can figure out how to do it, and they have abided with that policy for the past 50 years. The contents are in a nearby museum. The driveway to the Ming Tombs is lined with large statues of men and animals on the two sides.

The next day we rode 3.5 hours in an unheated bus to the Xinglong Observatory site, location of their 2.16m telescope and 5m LAMOST telescope. The latter, designed partly by Aden Meinel, has 4000 fibers to collect moderate-dispersion spectra of that many stars at one time. To date they have collected physical data (temperatures, metallicities, pressures) of 11 million stars.

The following day there were talks again and a show in the evening of acrobatics, ballet, magic, juggling, and balancing in a theatre south of Tiananmen Square. Another day we visited the Imperial Palace (Forbidden City), although most of the contents – everything that could be carried - were taken to Taiwan by Chiang Kai-shek, the nationalist Chinese leader. We also visited the Summer Palace. It has beautiful, colorful, mosaics in the long walkway along Kunming Lake and in the temples. In the evening the America delegation invited the ~30 Chinese hosts to an 18-course dinner in the Temple of Earth, including the largest shrimp I ever saw, sea cucumbers, fungus soup, preserved eggs, seaweed, etc., and ending with soup, cakes, and apples.

We then flew to Nanjing where we visited Purple Mountain Observatory; Zhang Yuzhe (84) was Director. He was trained in China and then went to the US where he obtained a Ph. D. at the University of Chicago under Dr. Van Biesbroeck. He returned to China in 1942 and was called "The Father of Chinese Astronomy". He died on 21 July 1986, a half year after I met him. While the other astronomers were looking at other telescopes on the mountain, Dr. Zhang asked me if I would like to see the Astrograph that he was using. I agreed; he had a car waiting to take himself, his wife, and me to see where he was doing research on asteroids. A Chinese stamp was printed in his honor. When I started an annual prize for astronomers in China, I named it for him.

The next day we visited the memorial and mausoleum for Sun Yet-sen, somewhat like that of Napoleon. We gave talks at Nanjing University. The Vice-President invited us to a 19-course dinner with five deserts, which took 2.7 hours. The next day we crossed the

Yangtze River on a 7km bridge built during 1960-1974 in the Cultural Revolution. I know one female astronomer who did part of the work. We went to the top of one tower; the statue of Mao that had been there had been moved to a room below.

I admire and am amazed by the discoveries and thinking of the ancient Chinese and am impressed by the recent progress in their infrastructure and science, although there are some aspects of China that I do not admire, just as there are aspects of America that I find regrettable. I love the natural beauty of the country and the heart-warming generosity of the people.

27.1 CHINA IN 1988-2017

There were 13 more trips to China in 1988-2017. Rather than describe each one, I will summarize them by topic.

Royal Tombs

Although one of the Ming Dynasty (1368-1644 AD) tombs had been opened, namely that of Emperor Ding Ling (1573-1620), the exciting ones undoubtedly are those of the Han Dynasty (221 BC -220 AD). The first Emperor and the one who united China for the first time was Qin Shi Huang Di (259-210 BC). The Chinese know the location of the tomb – a hill 76m = 250 ft high, which I have climbed – 22 miles east of Xi'an, it has remained unopened. Historical records say that there are crossbows and arrows primed to shoot anyone who enters the tomb. Palaces and towers were built within the tomb; they were filled with rare artifacts and treasures. There are rivers of mercury to simulate actual rivers.

In March 1974 some farmers were drilling a well 1.5 east of the tomb and discovered terra cotta statue fragments. By May 1974 archeologist started to excavate the site. What they discovered was undoubtedly one of the wonders of the world (Fig. 27-1) They found 8000 statutes in an area of 20,000 m². A large structure covering Pit 1 to cover the initial and future excavations was completed in 1979. A hundred wooden chariots were found. To stand at the entrance of Pit 1 (Fig. 27-1) and to look at that array of thousands of reconstructed statues is the most impressive view that I have ever seen in my life!

Little known to tourists are the excavations north of Xi'an for the sixth Han Dynasty Emperor, Jing Di, who lived during 188-141 BC and was Emperor during 157-141 BC. The square tomb structure has been untouched, but around the tomb are 81 burial pits, 20 and each side. Twenty of them have been opened. Several are covered with glass, so one can walk over them and see inside. Most of the 300,000 relics from the 20 pits have been moved to a new museum.

In this case, the statues were made of pottery with wooden arms and covered with cloth robes. But the arms and robes have disintegrated, so the formations of soldiers are armless and naked. Each soldier looks different in height, facial features, and hair style. From the genitals one can tell the males from the females and from the eunuchs (no testicles). Also buried were thousands of horses, cattle, pigs, dogs, sheep, goats, and articles for daily living. Few people and tourists in Xi'an know about this excavation and museum, so we saw only a couple dozen tourists there.

Chinese Poetry

I have not been very fond of most poetry. As a scientist, I prefer to state myself concisely and clearly, rather than by innuendo. I never liked the practice of forcing successive lines to rhyme, twisting the arrangement of words and thought to produce rhymes. Exceptions are the free-form poetry of Walt Whitman, Carl Sandburg, and John Dos Passos. Another exception is translations from foreign languages where there is no attempt to rhyme. Thus I learned about translations of Chinese poetry, of which the best are whose of the Tang Dynasty (608-953 AD). The best known poets, all living in the eight century, were Li Bai (701-762), Du Fu (712-770), Wang Wei (699-766), and Zhang Ji (730?-780?).

On one of my trips I saw a vertical scroll in Xi'an at the Forest of Stele gift shop, done in a regular script. The artist was one of the clerks. I bought the scroll (for $100) because I liked the calligraphy. Then a Chinese friend told me what was in the text. It was a poem by Li Bai called *Invitation to Wine*. Li Bai was basically an alcoholic who wrote many of his best poems, of which more than 1000 still exist, while he was drunk. The poem is a paean to poetry and wine.

In 2017 I was in the same shop in Xi'an and saw another scroll whose calligraphy I liked. The text is a poem called *Moored for the Night by the Maple Bridge* by Zhang Ji, another 8th century poet. The scroll was painted by Ho Zhengud and cost $350. The Maple Bridge is in the canal system of Suzhou. Li Bai said that that poem is better than all the poems that he, Li Bai, had written. School children are taught to memorize that poem. Wow, have I been lucky in buying Chinese scrolls!

Du Fu competes with Li Bai as the best poet of the 8th century, but Du Fu's poems are more concerned with the people, especially when families were separated by wars. Wang Wei wrote mostly about scenery. I encourage you to read some of those Tang Dynasty poems.

My favorite piece of music since I was a university student (60-70 years ago) has been *Das Lied von der Erde* (*The Song of the Earth*) by Gustav Mahler. It combines the intense romanticism of Mahler's songs with the mystery of Asian poetry. The songs were set to poems by Li Bai, Qian Qi, Meng Haoran and Wang Wei.

Transportation

All major cities in China are connected with fast trains. The normal trains, for which there are 129,000 km (80,000 mi.) of track, travel at 250 km/h (155 m/h). The high-speed trains (29,000 km = 18,000 mi.) travel at 350 km/h (220 m/h). There is a Maglev (magnetic levitation) train (Fig. 27-4) that connects Shanghai Pudong Airport with downtown Shanghai (30 km) that travels at 431 km h^{-1} (268 m/h). It was built by Siemens-ThyssenKrupp for a cost of $1.2b.

Fig. 27-4. The Maglev (magnetic levitation) train connecting Shanghai Pudong Airport with Shanghai. Its top speed is 431 km/h or 268 m/h. The fare is $8.

It is exceedingly smooth. Because of the marshy land, it rests on 70m long vertical columns every 25m.

One reason why China was able to build all those trains and tracks in the past two decades is because the government owns all the land, so obtaining rights-of-way was quick. Nevertheless, the infrastructure in China – and in many other countries - is decades ahead of the US.

One sees widespread construction and modern buildings everywhere, including in developing countries, but little in the US.

Jade Carvings

Jade is the hardest natural rock we know except for diamond. Nevertheless, from earliest times (before 3000 BC) Chinese have carved beautiful objects from jade, using diamond dust and abrasion. Also, the colors of jade objects are often beautiful. The tombs of royalty in the Shang Dynasty (1600-1046 BC) usually had three kinds of jade objects: (1) congs (pronounced chongs) are tubes that are square and highly decorated on the outside and with the circular hollow interior. The symbolic meaning of this design is that it represent the earth. (2) Symbolic hatchet heads; they are symbolic because they had dull rounded cutting edges. (3) Bi disks. See Section 33 about John Fountain for an astronomical use of bi disks.

One of my jade carvings is a green jade mountain 42 cm wide by 42 cm high (13x13 in.) obtained in Hong Kong after much bargaining for $1000; the initial price was $4000. Others are smaller but more intricate. I was fortunate in buying those before there were enough rich Chinese to drive the prices out of my financial range.

My favorite jade carving is a five-tiered cong that is 9.5 inches high by 3.5 inches square (24 x 8.9 cm). We bought it after hard bargaining for $1200 in Macau, when it was still a Portuguese colony. The initial price was $5000. It is the largest cong I have ever seen in museums or in books. John and I thought that I. M. Pei should have used the design for an office or apartment building with a large central atrium. The trip there from Hong Kong was in a hydrofoil. In 2018 the Chinese completed a 55 km bridge overseas across the Pearl River estuary from Macau to Hong Kong, at a cost of $18.8b.

Chinese Opera

Chinese opera is very different than western opera. It has a smaller orchestra and music on a pentatonic scale, but much more elaborate costumes and more acting in the form of acrobatics. A screen above or to the right of the stage gives the words being sung, both in Mandarin and English. My favorite Chinese opera actor is the late Mei Lanfang (1894-1961), who played the female roles (no females played originally). I have videos of all of his 50 operas. An astronomer who loves Chinese opera is Xiang-Tao He, Astronomy Professor at Beijing Normal University. On nearly every one of my trips to China I gave talks to his astronomy classes and he took me to performances of Chinese opera. The last time it was held in the new National Centre for the Performing Arts in a spectacular building. From a distance it looks like a dome set in the middle of a circular lake. On one side there is a broad stairway down under the lake to come up into the dome. It was designed by French architect Paul Andreu. Why doesn't the US have interesting new architecture (except for the buildings of Frank Gehry) as all the foreign countries do? We are falling behind the rest of the world in so many different ways.

Stamps

Even during my first trip (1985) to China I was attracted to the beautiful designs and wide portrayal of Chinese life on their postal stamps. They were very cheap when letter postage within the country was a few cents. I soon started to collect them. From my childhood experience with collecting stamps, I knew that one should only collect mint stamps and never lick the backs. At a time in Chinese history when people had little money, saving in banks was unreliable and stocks did not exist in a communistic economy, people started to collect stamps. That lasted until 1992 when some Chinese were rich enough to compete with me. By about 2010 I lost interest in favor of other things, so

I sold my stamps from before 1992 for $18,000. I still have many from after that date but they are only worth their face value.

Tibet

In 2017 I was again invited by Gang Zhao, Director of the Chinese National Observatory, to visit China, to give talks on topics of my choice, and to tour some sites. My travel, hotel costs, and per-diem would be paid. Such generosity! That year I asked to visit Tibet with my friend Yao Huang. It was supposed to happen after my visit to Beijing, but that would have placed the Tibetan part during the 19 October annual meeting of the Chinese Communistic Party Central Committee in Beijing. The government feared having protests in Tibet during the meeting and forbid any foreigners in Tibet then. Therefore Yao and I flew to Lhasa by plane before the meeting.

The flight from Beijing to Lhasa took 4.5 hours on an Airbus 319. Gongko Airport is 62 km south of Lhasa. The road goes over a bridge 3.5 km long and built in 2016. Then a tunnel 3.5 km long. Lhasa has a population of 1 m (partly Tibetan, partly Han) out of a Tibetan population of 3.2 m. As one can expect for a small population spread out over such a large region, there are many Tibeto-Burman languages that are mutually unintelligible. The largest population majority is Bhuddist, who ran the country until the Chinese took over politically in the 1950s.

One gets a very different viewpoint of Tibet by visiting it. Before the take-over, citizens received only several years of schooling – only until they could read the Buddhist scriptures. There was no teaching of science, math, history, geography, or languages. Slavery existed until 1958 when the Chinese abolished it. Many families were so poor that they had to sell one child into slavery to get enough money for food. After the take-over, schools were free for K-12 and the new colleges were free of tuition. Medical help was free for all. In schools, the children are now taught Tibetan, Mandarin, English, science, math, history, geography, etc. For the first time, Tibetans learned about the rest of the world. Lhasa is thriving with zero poverty. Of course, many of the best positions were taken by Han from China. On the train many Han Chinese were traveling to and from Tibet. We met a Tibetan couple (he had been a school principal) who were taking their 10-year old grandson to Shanghai for the first time. The reason why we receive a different viewpoint in the west is because the Buddhist are no longer running the country, although most of the temples are open for worship; religious people in the west regret that loss. But the Buddhists had kept the people in suppression and ignorance, just as the Inquisition in 15th century Europe did.

We stayed at the Intercontinental Lhasa Paradise Hotel, a 5-star hotel, the most luxurious hotel in which I have ever stayed. You should see the ~100 choices in their buffet meals. When registering, a doctor checked my blood oxygen level; it was 85%, so they installed an oxygen generator in my bedroom. I used it at night but not during the day. The elevation of Lhasa is 12,000 ft. I questioned being put up in such an expensive hotel, but was told that the tourist agency that the Chinese National Observatory used liked to increase their commissions. The cost for the hotel, meals, guides, driver, minivan and admission charges, was $1800 for two

We went to the Potola, the stone palace for the Dalai Lama that has about 1,000 rooms, but no tourists were allowed in. The red brick part was built in the 7th century and the white stone part in the 17th century. The Summer Palace downhill was built for the 14th Dalai Lama but he used it only for two years before he moved to India at age 24. We toured the inside; monks were sitting around and drinking tea. Except for the elaborate decorations, I would have preferred a Motel-6 for comfort. Lhasa has modern buildings of up to 10 stories. We visited one temple and Yao visited another, but I asked to go to the hotel because I was feeling weak; my blood oxygen was down to 80%. I slept 3 hours while Yao went out to buy hamburgers.

The next day we took the train from Lhasa across eastern Tibet through Gilmud to Xi'an (Fig. 27-6). It took 33 hours, rose to 16,000 ft., and had been completed in 1984. The train was not pressurized but it had oxygen at the head of each bunk; rooms had two double bunks each. The train has about 15 cars long including a dining car. But the scenery was fantastic! Grassy valleys (it was all above timberline) with yak herds, surrounded by snow-capped peaks and lakes. We saw the lake that was the source of the Yangtze River. We had a dinner of beef, green peppers, chicken, rice and soup for $15 for two.

We changed trains at Xiniga. We gradually became acquainted with all the people in our car. We shared a room with a man and his wife who has a chain of Peking Duck Restaurants. He shared with me his Peking Duck meals. We followed the Yellow River to Xi'an.

Fig. 27-5. The train from Llasha to Xi'an across eastern Tibet going up to 16,000 ft (4900m). It had 15 cars including a dining car. Each bed had oxygen.

28

EARLY CHINESE DISCOVERIES

We have all heard that the Chinese discovered gunpowder, fireworks, kites, and other things years ago, but before the work of Joseph Needham, we did not realize the extent of Chinese knowledge in science, math, agriculture, physics, medicine, and other fields. We did not know before Needham that in the first century BC the Chinese were drilling wells for brine-salt 4800 ft. (1500 m) deep. When oil was discovered in Pennsylvania in 1859 a few meters below the surface, they had to import Chinese engineers to show them how to drill short wells. Or that the Chinese were using the decimal system in the 14th century BC!

Joseph Needham (1900-1995) was a Cambridge, England Professor of biochemistry who became acquainted with Chinese scientists visiting Cambridge University in the 1930s. He was sent to China in 1942 as a British science advisor at the British Embassy. He stayed there through WWII and for two years more as an UNESCO advisor. He learned to speak and read Chinese. Colleagues in China called his attention to some ancient Chinese books. He realized that the Chinese discovered more about science and math 2000 years ago than was known to Europeans until the 17th or later centuries. He spent the rest of his long life writing 58 books in a series called *Science and Civilization in China*. These are too much for me and others to read and digest, but fortunately Robert Temple summarized Needham's discoveries, with his permission, in a book called *"The Genius of China"* (London: Prion Books Ltd., 1998). Let me tell about some of those discoveries. Much of the rest of this section has been taken from Temple's book.

In the west we think that sunspots were discovered by Galileo in the 17th century when he received a Dutch telescope. But the Chinese first discovered sunspots, using pinhole cameras, in the fourth century BC. (the Arabs discovered them in the ninth century AD). Needham found 112 references to sunspots between 28 BC and 1638 AD; there are probably many more. The Chinese found the solar rotation period of about 28 days and the sunspot cycle of about 13 years.

Western astronomers discovered the solar wind in the 20th century, but the Chinese discovered it in the 6th century AD. They were avid observers of the sky (for reasons of astrology). They recorded the occurrence of hundreds of comets and European astronomers were able to use those observations to compute the orbits of dozens of long-period comets. But the Chinese noticed that the comet tails always pointed away from the Sun, so they surmised that the Sun was emitting particles that forced the comet tails to point away from the Sun.

Please see Section 2 for the Chinese observations of the Crab Nebula of 1054 AD. There are still 59 unidentified novae and supernovae for which Lundmark (1921) collected Chinese observations.

The Chinese used the decimal system at least as early as the 14th century BC, but probably much earlier. It came about naturally because of their use of the abacus, which has 10 beads per column. In the west the first use of the decimal has been traced to a Spanish manuscript from 976 AD.

The use of a zero in calculations came later. The Chinese initially used a blank space, e.g. 405 was written "four blank five", starting in the 4th century BC. The use of a symbol for zero may have started first in India or Indonesia. The use of negative numbers started in the 2nd century BC in China. To us who measure temperatures in the Fahrenheit or Centigrade scales or who deal with bank accounts, negative numbers make sense. The Chinese used decimal factions, square roots, and higher roots as early as the 1st century BC

However, the early practical Chinese developments in agriculture and engineering had greater social importance. For instance, the Europeans placed harnesses around the necks of horses, which tended to choke the horses. As a result, a single horse could only pull 250 kg. Although in Roman times southern Italy could produce plenty of grain to feed the Romans, it was easier to get grain from Egypt by ship than to transport the grain from southern Italy by horse-drawn vehicles. The Chinese, who sometimes used people to pull plows, realized that the harnesses should be placed around the chests of horses or oxen, not around their necks. With Chinese harnesses single horses can pull 1500 kg.

European cultivation was primitive. Farmers scattered seeds on the ground, wasting most of them. The Chinese planted seeds in rows after turning over the soil with iron plowshares. The Chinese learned to make cast iron in the 4th century BC. They had clays that would stand temperatures up to 1100 C. With coal as a fuel and a mixture of 6% phosphorus, in the 4th century BC they produced temperatures high enough to melt iron ore. The Chinese also used iron hoes to remove weeds from the rows so the water went to the crops. Their crop efficiency 2000 years ago was several times higher than European crops in the Middle Ages!

The Chinese were the first to make steel from (brittle) cast iron in the 2nd century BC. The trick was to reduce the amount of carbon from the 4.5% in cast iron. That was not widespread in the west until 1856 when Henry Bessemer developed the Bessemer steel process, although William Kelly in Kentucky did it in 1845 by bringing in Chinese experts.

In ancient times salt could be obtained in eastern China by drying sea water near the coasts. However, transportation several thousand kilometers to western China was not practical. Instead, in western China they drilled into the earth to bring up brine. Even today 16% of the salt used in western China comes from brine wells. What the Chinese did as early as the 1st century BC was drill down until they found brine. They first dug through the ground until they reached rock. Then they built towers and dropped cast iron bits, using hemp ropes, repeatedly to work their way through the rock. The bits were up to 3 meters long and weighed up to 100 kg. Then they lined the holes with bamboo tubes, which have the tensile strength of 25 tons per square cm. The drilling went down from 10 cm to a couple meters per day. The brine was pumped out using double-acting piston bellows, developed in the 4th century BC. The average well was 1000 m deep but the record was 1500 m. An imperial edict limited the number of wells to 1069. Soon the government wanted to declare a monopoly on the wells. Using spies, the owners of private wells simply dismantled the derricks and spread some dirt over the holes and trays when governmental inspectors came around.

Once the brine was brought to the surface, it was pumped into large cast-iron trays that were heated from below to evaporate the water. Up to 5100 salt pans were used with a single well. Furthermore, some wells encountered methane gas, so that was brought up as a fuel to heat the trays. By 1834 the Chinese techniques were brought to Europe and in 1859 it was used for the first oil well in the US.

The Chinese built the first suspension bridges in the 1st century AD, crank handles in the 2nd century BC, and chain drives, like on bicycles, in the 10th century AD. A clever technique for underwater salvage occurred in the 11th century AD. In 1064 a pontoon bridge across the Yellow River was washed out by a major flood. The series of boats holding the roadway across the river had been held with cast iron chains attached to giant iron figures on the banks, which were pulled into the water. An engineer had several boats filled with sand and attached them to the iron figures in the water. Then they shoveled out the sand, the boats rose, and lifted the figures up and to the banks.

The Chinese developed the first lacquer in the 13th century BC., the wheel barrow in the 1st century BC, the fishing reel in the 3rd century

AD, the stirrup in the 3rd century AD, porcelain in the 3rd century AD, and matches in the 6th century AD. They developed wine by the 2nd century BC, and brandy and whiskey in the 7th century AD.

We in the western world believe that the circulation of the blood was discovered by William Harvey in 1628. However, the Chinese knew in the 2nd century BC that the blood circulated and that it was the heart that did the pumping. They computed that the blood flowed 2500 m per day.

The Chinese discovered the cures for deficiency diseases, like beriberi, scurvy and rickets. For instance, Han Yű (762-824 AD) noticed that beriberi, that we now know is caused by the deficiency of vitamin B_1, was more prevalent in southern China, where the people ate washed rice; in northern China people left the husks on the rice.

By the 12th century AD the Chinese discovered that the scurvy among sailors on long voyages was due to the lack of fruits, primarily citrus fruits. In their long voyages to as far away as the east coast of Africa, using dozens or hundreds of ships, they grew orange and lemon trees aboard some ships to give the sailors what we now call vitamin C.

The Chinese were acquainted with diabetes as early as the 6th century AD. They were familiar with goiter, which we know to be due to a lack of iodine, and treated it successfully with Sargassum seaweed in the 1st century BC. The Europeans did not catch on to that treatment, from the Chinese, until the 12th century AD. However, the most surprising Chinese discovery is the use of inoculations to avoid smallpox (syphilis was called the "grand pox"). It is not clear when they started to use the technique, certainly by the 10th century AD, which involved sticking a swab of cotton into the nose and let a small dose of the virus be breathed in through the membrane. The samples were obtained from scabs from diseased people but kept first for one month so that the viruses were dead.

In the fields of transportation, the Chinese built kites as early as the 5th century BC, but kites were not mentioned in Europe until 1589.

Some kites were large enough in the fourth century BC to carry men. They built parachutes in the 2nd century BC – long before Leonardo de Vinci sketched parachutes in the 16th century. The Chinese built rudders in the first century AD, when the Europeans were using steering oars on small ships. The Europeans learned about rudders from the Chinese in the 12th century AD. The Chinese built multiple sails on ships and used bamboo to stabilize them. They built the first watertight compartments in ships in the 2nd century AD, which the Americans developed in the 18th century. They built the first pound locks on canals in the 10th century AD. Please see Robert Temple's book for many more inventions that the Chinese made 1000-2000 years before the westerners.

The young Chinese do not know much about these ancient Chinese discoveries because of the education gap during the Cultural Revolution and that the Chinese government likes to emphasize the progress made since 1949. Therefore I have often given talks to Chinese students about what their ancestors did.

29

BLUE STRAGGLERS

Blue stragglers are stars in open clusters that are above the main sequence turn-off points. Why are they still on the main sequence when less massive stars have already exhausted the hydrogen in their cores? Explanations tried were that they were formed later, that they were not members of the clusters (foreground or background stars), or that they were returnees from the giant regions. Observations showed that none of those were valid.

The most widely explanation accepted in the 1965-1985 was that they were members of double-star systems and the evolving companions were dumping hydrogen-rich material onto the blue stragglers, allowing them keep converting hydrogen to helium in their cores (McCrea 1964). There were two problems with this explanation. One is that observations by several people failed to show that most blue stragglers were members of double-star systems. The second is that even if hydrogen was dumped onto the outer layers of the blue stragglers, how did it get into the cores? Hence the double-star explanation failed.

Observations by Pendl & Seggewiss (1973) and Mermilliod (1982) showed that many blue stragglers have abnormal spectra such as Ap, Am, Be, Of, etc. I obtained spectra of 16 blue stragglers (Abt 1985). For the stars in clusters of large ages $10^{8.3} - 10^9$ yr, 62% were Ap stars, mostly Ap(Si) stars with low rotational velocities. This confirmed what Pendl & Seggewiss found. These are the kind of stars found by Babcock (1958) to have strong magnetic fields (~ 10^3 gauss). Schüssler & Pähler (1978) and Hubbard & Dearborn (1980) found that strong magnetic fields in B4-A5 stars are sufficient to cause magnetic mixing of the interiors of stars; that would bring hydrogen-rich material from the outer layers into the cores.

What about the blue stragglers in the intermediate-aged clusters that are not Ap stars? Abt (1979) found that it takes 10^8 yr to produce Ap(Sr,Cr) spectra, so those stars could be ones who will be Ap stars, although strong internal magnetic fields remain regardless of what happens to the surface layers.

Mermilliod found that the blue stragglers in young clusters (earliest stars of types O6-B2) have unusually broad lines. He and I found mean projected rotational velocities $V \sin i = 220$ km s^{-1}, compared with 150 km/s for field O6-B2 stars (Slettebak 1970). Also of 15 such blue stragglers in young clusters, half have emission lines, which is an indicator of rapid rotation. So do such stars have rotational mixing to bring outer hydrogren into the cores? Tassoul (1978) does not rule that out in massive stars, but admits that there are too many unknown parameters, such as the core rotation relative to the envelope rotation.

Therefore we have two mechanisms that can replenish the amount of hydrogen in the cores of blue stragglers and keep them on the main sequence longer: magnetic mixing and rotational mixing.

30

JOHN FOUNTAIN

John Fountain was born and raised in Quincy, IL, across the river from Hannibal, MO. His father had disappeared when John was young. In his teens John was developing allergy or lung problems, so he and his mother moved to Tucson, where John went to Catalina High School and the Univ. of Arizona. After graduating from the Astronomy Dept., he applied for a position at the Lunar & Planetary Laboratory. He told me that he had a 30-min. interview scheduled with Director Gerard P. Kuiper. Kuiper spent the 30 min. telling John of his plans for the laboratory. At the end of his monologue he said "I'm sure that you will be happy here."

John worked for Kuiper for 15 years, being his primary research associate because John became an expert in photographic techniques and how to interpret photos. They found at least one new satellite of Uranus. Then Kuiper died of a heart attack in Mexico. John worked for Bradford A. Smith, who then went to Hawaii. At that stage (1991, I believe) he checked his finances and decided that he could retire if he lived carefully. He had developed an interest in astroarcheology – what ancient people knew about astronomy. I will tell about his major projects below.

I met John in the 1960s when his university class toured Kitt Peak and I had lunch with the class. We immediately became friends. When I had a bypass surgery in 1996, I spent four days in the hospital and then was sent home with the instructions to do no physical work. I read books. John offered to stay with me, do the cooking, shopping, keeping up my house, and walk our dogs. We got along very well. Even our dogs got along. At the end of the month I suggested that he stay, and he did for 20 years until his death. He sold his house. He had architectural plans to build a house in Torrey, Utah and had purchased land there, but his health was failing and I talked him into not leaving Tucson with our house close to Tucson Medical Center. He had inherited a rare lung disease that ran in the male side of his family. He went frequently on field trips for his research and I took him on trips abroad. In 20 years we never had an argument. Those were the best 20 years of my life.

One of the projects that we did together was to write up the results on the Crab Nebula Supernova. That was discussed in Sec. 2.

Another published paper concerned Chinese bi disks from the Shang Dynasty (1600-1068 BC). Royalty of that era were buried with three kinds of jade carvings. One type was congs that were said to symbolize heaven and earth. The second was ceremonial hatchet heads, which had dull cutting edges. The third were bi disks: flat round disks with notches cut into their edges; they were thought to be used for astronomical reasons. We found that if they were held toward the north pole, the notches could align with certain bright stars around the pole. John had a program that showed the positions of those stars at different times. He found that bright stars fit into the notches of the bi disk at the Museum of Science and Industry in Chicago only during the years 1700-1200 BC.

Then John became interested in geoglyphs. Those are large designs on the ground in the US southwest – either by rocks laid in lines or by cutting through the dark desert varnish. One thinks of the designs on the plains of Nazca in southern Peru. The ones in California and Arizona were 10-20 meters in size and had geometric or animal patterns. A few were known, but John found many more.

Another research project of John's was summit trails. There are many small (~100 m) conical volcanic hills in Arizona and Sonora. John found that many had trails cut into them that went straight up to the tops. Those are similar to the rock structures in Yucatan, e.g. Chichen Itza, Tikal, and Palenque. But which came first? Did the American Indians copy the Mayan ruins or the other way around? John studied the azimuths of the trails – they tended to be on the north sides of the hills. He gave a paper on those at one of the Rock Art conferences in San Diego, but the papers from that conference were never published. I am having difficulty reconstructing that paper.

John and I enjoyed our trips abroad because we both have wide interests and are curious about many things. He was an excellent traveling companion. Most of the trips were because I attended scientific meetings (e. g. Brussels, Paris, South Korea, China, Thailand), but others were for fun (e.g. Rome, Cambodia). Also, we both enjoyed concerts, plays (especially John), classical music and reading (especially me), and mystery stories on BBC TV. He was an excellent gardener. When he had a house on 7th St., he had a small greenhouse and grew orchids – so many that he regularly sold them to supermarkets.

However, his lung disease gradually became worse. First, he had to use an oxygen generator when he was in our house. Then he had to carry a tank of oxygen with him whenever he left the house. Our last trip together was to Lijiang, China, where I had a meeting. See Figure 30-1.

Fig. 30-1 John in April 2011 in Lijiang, China. Behind him is Jade Garden Snow Mountain (5,596m = 18,360 ft.). The Yangzi River flows from Tibet and hits the north side of this mountain, sending it east across China to the Pacific at Shanghai. If this mountain had not been in the way, it would flow south to Vietnam and central China would be a desert. (Photo by the author)

On my birthday, 26 May 2016, he felt unwell and wanted to go to the hospital. I immediately started to bring my car closer to our front door, but he asked that I call an ambulance. That is the last time I talked with him. By the time I reached the emergency ward, he was sedated with tubes to help him breath, keep his heart going, etc. For 12 days he gradually declined. Then the doctor told me that his lungs were failing (pneumonia), his heart was failing, as were his kidneys and bone marrow. They could not make him recover, so I had to give permission to pull the plug. He died with me holding his hand.

31

THE DIFFERENCE BETWEEN METAL-RICH AND METAL-POOR STARS

There had been many papers written in the past on whether metal-rich (or Population I) stars and metal-poor (or Population II) stars have differences in their frequencies of binaries. The published differences depended on the resolutions of the spectrographs used. If one used a low-resolution, one was able to detect only the short-period binaries and if one used a high resolution and observations over long times, one could detect the long-period binaries.

The problem may have been resolved in Abt (2008b). Reviewing only studies based on high-resolution spectra (Latham et al. 2002, Preston & Sneden 2000), it became obvious that the metal-rich stars had frequencies of binaries that peaked at 30 days while the period distributions of the metal-poor stars peaked at 1000 days, i.e. they lacked short-period binaries. The explanation for this was that Aarseth & Hills (1972) and Kroupa (1995) showed that most binaries are formed by capture in dense clusters. The longer they stay in those clusters, the closer the binaries become. Kroupa calculated that the halo stars that Latham et al. observed are ejectees from globular clusters or the remains of disrupted clusters, such as Palomar 5. If that is true, the metal-poor stars may not have remained in globular clusters long enough to form short-period binaries. In contrast, the metal-rich stars from open clusters stayed in those clusters for 10^8- 10^9 yr. Preston & Sneden proposed a different scenario. It would be interesting to know the distribution of periods for similar stars in globular clusters.

32

LOCAL INTERSTELLAR BUBBLE

The Local Interstellar Bubble (or Cavity) is an irregular region from 50 to 150 parsecs in radius (160 to 500 light years) from the Sun. Its density is 1/100 to 1/1000 that outside the bubble. See Fig. 32.1. It was swept nearly clean of gas by several supernovae. Using data from double stars, single stars, open clusters, emission regions, X-ray stars, planetary nebulae, and pulsars, I established (Abt 2011) that it has three sub-regions. The region toward the galactic center has stars as early as O9 V with ages of 2-4 million years and a pulsar (PSRJ1856-3754) with a spin-down age of 3.76 million years. The central lobe has stars as early as B7 V and therefore an age of about 160 million years. The Pleiades lobe has stars as early as B3 and therefore an age of about 50 million years. The gas density is about 0.05 atoms per cubic cm (0.8 atoms inch^{-3}), compared with a hundred times that outside the bubble. In the solar system the density is 10^{13} atoms per cubic cm. We know nothing about the supernovae that caused this bubble.

Phillips & Clegg (1992) and Smith et al. (2007) derived a temperature of 1-2 million deg. in the bubble. Walsh & Lallement (2008) found five stars at the perimeter of the bubble (in my maps) with lines of O VI 1032Å and C II 1036 Å, confirming the high temperature. The Sun went into this region long after the supernovae exploded and will stay inside the bubble for thousands of years more.

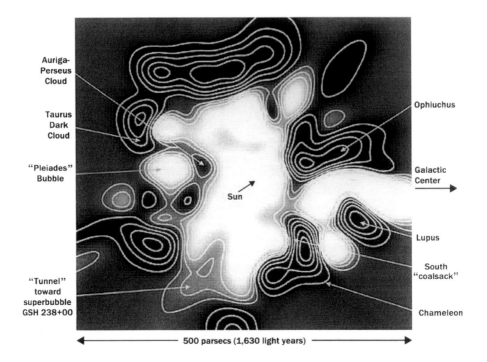

Fig. 32-1. The Local Interstellar Bubble is an irregular region about 400 parsec (1300 light years) in diameter in which three supernovae drove 99+% of the interstellar gas away. The Sun drifted into the bubble long after the supernovae exploded. The lobe toward the galactic center (to the right) was produced by a supernova that left pulsar PSR1856-3754, age 3.76 m years. The lobe around the Pleiades (left) occurred about 50 m years ago. The main part of the bubble was produced by supernova about 160 m years ago. The Sun and solar system will stay in the bubble for thousands of years more. (Photo by the author.)

The radio astronomy observations showed the shape of the bubble and it central density and temperature. My contribution was to determine, with MK classifications, the ages of the supernovae that caused the bubble. The bubble is a very low-density region of interstellar space with a very high temperature, due to three supernovae with ages of about 4, 50, and 160 million years. This was my most exciting project in the past decade, but few people have discovered this paper

33

SPAIN

Drs. Jose Docobo and Josephine F. Ling at the Observatory Ramon Maria Aller in the University of Santiago de Compostella (established in the 16th century) are very active researchers of visual double stars. For the IAU they publish and distribute semi-periodic research notes giving new measurements and orbital elements obtained internationally by astronomers. They have also organized several international conferences on double stars. I attended three of those because I always had new results on double stars. One, in 1996, was held in connection with the 500th anniversary of the University of Santiago de Compostella.

Travelling to Santiago de Compostella in northwest Spain (Gallicia Provence), always meant flying to Madrid and changing planes. I couldn't resist staying a day or two in Madrid to see Picasso's *Guernica* at the Museo Reina Sofia. Also to the Prado to see paintings by Velazquez and Juan Miro. Also I loved seeing the Flamenco dancers in evening performances.

Santiago de Compostella, just north of Portugal, is the site of the cathedral that claims to have the bones of Saint James. They are in a coffin below the altar. At least that is what a bishop in 813 AD said. A favorite hike in Europe is to start in southern France just north of the Pyrenees, to hike westward across the Bosque region of Spain, and ending at the cathedral to see the coffin of St. James. Then the pilgrims receive certificates.

In the Cathedral we saw a demonstration of the Botafumeiro. That is a brass & bronze (silver coated) censer that is 1.6m high and weighs 80 kg. It contains 40 kg of charcoal and incense that is swung by eight citizens from a 21m rope attached to the ceiling. After 17 swings (80 sec. each) out over the audience it reaches an angle of 82°.

On the trip in 2011 to a workshop on double stars in Santiago de Compestela, I decided to stop in Bilbao in northern Spain to see the Guggenheim Museum, designed by Frank Gehry (Fig. 33-1). Bilbao (350,000 people) is in a pleasant valley in Bosque country - the kind of place where one would like to retire. The museum (cost $89m) was dedicated by King Juan Carlos on 18 Oct. 1997. The outer skin of the building consists of thousands of titanium sheets, 0.04 mm thick and made in Pittsburgh. It has 11,000 m^2 of exhibition space. Inside are exhibits by Robert Serra and Jeff Koons (e.g. *Tulips*). Outside is Jeff Koons' *Puppy*, 12 m high. It is made of steel and covered with beds of pansies. What a light humorous touch to people who think that art should be something serious and conservative!

Some of the participants of the workshop were from Russia (Yury Balega, Oleg Malkov, N. Melikian), the US (Bill Hartkopf, Brian Mason), and hosts Jose Docobo and Josephine Ling. I talked on the age of the Local Interstellar Bubble. Bill Hartkopf talked about Future Changes to the Washington Double Star Catalog.

Fig- 33.1. The Guggenheim Museum in Bilbao, Spain, designed by Frank Gehry

34

JORDAN AND PETRA

In December 1997 I received an invitation from Dr. Hussein Rashid of Mu'tah University in Mu'tah (south of the capital Amman, Jordan) to attend an international conference on astronomy during 4-6 May 1998 with a side trip to Petra, all expenses paid. The university was established in 1981 and has 17,000 students. I agreed. Several months later I heard from President Mohammad Adnan Al-Bakhit of the Institute of Astronomy and Space Sciences at Al al-Bayt Univ. in Mafraq (45 km northeast of Amman). That university had 4,500 students (now 18,000) and was established in 1992. That is where the conference would be held. The attendees were astronomers from Saudi Arabia (1), Iran (1), Egypt (2), Tunisia (2), Algeria (2), Jordan (15), France (2), England (3), Mexico (1), and the US (3). I had submitted abstracts for three talks. The talks by the others ranged from geophysics to cosmology.

The airline tickets arrived on April 28th (the day after Mother's last birthday) for flights starting on May 2nd. They involved flights from Tucson to Denver to NY (LaGuardia) to Amsterdam and to Amman on Royal Jordanian Airlines. The meals on the airlines and in the whole conference were mostly chicken.

On May 4th after introductory talks and talks on solar and polar winds, we were taken to the 2-3rd century AD Roman ruins at Umm Qais (see Ancient Gadara Umm Qais on Wikipedia), including a Roman theatre. Then we went to northern Jordan where we looked west across the Jordan River and Lake Tiberius (Sea of Galilee) into Israel and north across the Yarmouk River to the Golan Heights. The Golan Heights is a mount (2800 m high) of 1800 km² that was captured by Israel from Syria in the 1967 Six-day War. We could see military vehicles traveling and people walking on the top.

The next day I gave a talk on binary stars, along with talks by others on propulsion systems, cosmology, and teaching. In the afternoon we went to Umm el-Jimal, a 1st century ruins 17 km east of Mafraq and 10 km south of Syria. The ruins date back to Roman, Byzantium, Nabataean, Druze, and Bedouin cultures. They had city walls with four gates, an aqueduct water system, and 159 houses. It gave me an eerie impression.

The third day I gate a talk on the lifetimes of papers in various sciences and chaired the last session. There were plans to start an Arabic astronomy journal (and I submitted a paper), but that hasn't happened yet. In the afternoon we visited Jerash, just north of Amman. It has been inhabited since the Bronze Age, then by the Greeks, and then by the Romans. The ruins are said to be the best ones from Roman times. There were hundreds of limestone columns, a Roman theatre seating 3,000 people, grain storage chambers, the Temple of Artemis, and (from Byzantium times), the Basilica of St. Theodore and two other churches. There was a Nymphaeum: a pool of water 600 m² in area and 3 m deep. One could see in buildings the crude blocks from Greek times, the precise blocks from Roman times, and the smaller blocks from the Bedouin times.

The next day we went to Petra. Driving there in a bus, we passed the most arid and featureless desert I have ever seen, but rich in potash. We passed the Hejaz railway, which Lawrence of Arabia blew up during WWI. We (13 people) hired a guide for a four-hour hike into the ruins. On the way in along the 1.2 km Siq we passed on the side wall a channel cut into the sandstone for water to flow. At the intersection of channels is the famous Treasury, shown in many pictures. It is an elaborate building cut into the sandstone, but was actually, despite its recent name, a place where dead people were prepared for burial.

To carve it they started at the top and worked downward. In an open village area there was a theatre, a colonnaded street of shops, and a Nabataean church, the only free-standing building. In the surrounding area wherever there were walls, buildings were cut into the sandstone. The whole area (about 15 km^2) was inhabited since 7000 BC or earlier and was the Nabataean capital since the 4th century BC. It was on the main land route from Asia to Europe until a sea route was developed. The people became experts in water harvesting, desert agriculture, and stone carving. The 20,000 Nabataean people fell to the Romans in 106 AD. The ruins were discovered in 1812 by Johann Ludwig Burkhardt. It is now one of the Seven New Wonders of the World and a UNESCO Cultural Heritage Site; it attracts 1 million people a year.

I recommend the large picture book named *Petra* by Jane Taylor with a forward by King Hussein, 1993 (London: Aurum Press Ltd.).

I stayed in Jordan until Sunday May 10, enduring the Mullah's calls to prayers from a minaret every two hours from 5 am. I talked with prolific Iraqi astronomer Hamid M. K. Al-Naimiy whose observatory had been bombed in the Iraq war because the domes looked like military installations. Fortunately he had removed the optics before. During a trip abroad to a meeting, he escaped to Syria and Jordan; he was a Professor at Al al-Bayt University in Jordan. Later he moved to the Sharjah University in the United Arab Emirates and has been the Dean of Sciences since 2006. He would like to build a 1.5m telescope.

35

GEORGE ELLERY HALE

The person who did the most, by far, to organize national and international astronomy and bring it into the 20th century as astrophysics, a term he coined, was George Ellery Hale (1868-1938). He was born in Chicago; his father started a company that built elevators, much needed after the Great Chicago fire of 8-10 Oct. 1871. He was educated at MIT, Harvard, and the Berlin Academy. He was a professor at the University of Chicago from 1893-1907.

Hale started the Yerkes Observatory in 1893 with a 40-inch (1m) refractor, then the world's largest telescope. It was exhibited at the 1893 Colombian Exhibition in Chicago before being moved to a site on Lake Geneva in southern Wisconsin. Then Hale realized that California had better weather for observing and he started the Mt. Wilson Observatory with a 60-inch (1.5 m) reflector and a 100-inch (2.5 m) reflector. He brought in astronomers such as Harlow Shapley and Edwin Hubble. He also started the two solar telescopes on Mt. Wilson. With one he discovered that sunspots have strong magnetic fields, by using the Zeeman Effect. Before WWII he started the Palomar Observatory, farther than Mt. Wilson from the Los Angeles city lights. The Palomar 200-inch (5.1 m), belonging to the California Institute of Technology (Caltech), was completed in 1948. He helped convert the Troop Institute of Technology into the California Institute of Technology (Caltech). Caltech started an Astronomy Department in 1948.

In 1895 Hale started *The Astrophysical Journal, An International Review of Spectroscopy and Astronomical Physics*, owned and operated by the University of Chicago Press. He realized the astronomical research had to be an international endeavor, partly because of its need for round-the-clock observations. After organizing international astronomers at the St. Louis Exposition in 1904, he started the International Union for Cooperation in Solar Research in 1907, which in 1919 was broadened to become the International Astronomical Union (IAU). During WWI he organized the National Research Council, a governmental organization to advise the US government on scientific matters.

The goals of the IAU are to promote and safeguard astronomy, including research, education, and international cooperation. The individual members have Ph.D.'s or beyond. They come from 82 countries. The IAU cooperates with other scientific organizations.

36
THAILAND

One of IAU's activities is the International School for Young Astronomers. Every three years there is a school in some part of the world, mostly in developing countries, to train students in that region with lectures and workshops in order to educate young people in astronomy.

In 2001 I was invited to be one of 10 lecturers for an IAU-UNESCO International School for Young Astronomers to be held in Chiang Mai, Thailand. Several of the active arrangers of these schools has been Kam-Chiang Leung, born in Hong Kong, trained in the US and Canada, and a Professor at the University of Nebraska, and Michele Gerbaldi from France. The nine astronomers who came to Chiang Mai (Ed Guinan dropped out for medical reasons) included Kam Leung (USA), Michele Gerbaldi (France), Boonraksar Soonthorthun (Thailand), Jean-Pierre De Greve (Belgium), Kwing L. Chan (China), Kwang-Sang Cheng (China), Ajit Kembhavi (India, Vice President of the IAU), Don Kurtz (England), and myself. Don had worked for many years in South Africa but after the termination of apartheid, the crime rate became so bad that he moved to Central Lancaster University (formerly Preston Univ.) in England. The 36 students included 17 from Thailand and 19 from other southeast Asian countries. They were working either on B.S., M.S, or Ph.D. degrees.

In the two months before going to Thailand I prepared nine lectures on stellar rotation, stellar classification, radial velocities, visual binaries, spectroscopic binaries, stellar masses, spectrographs, stellar chromospheres, and *How to Write a Paper and Get It Published*. Before the workshop the lecturers had discussed the topics that needed to be taught and assigned them to the various lecturers. There are also talks by the students and the students did observing on Doi Inthanom.

I won't write about the many Buddhist temples that we visited, the orchid farm, the lacquerware and celadon factories, and museums, but I cannot refrain from mentioning the wonderful fruits that are available in tropical countries but generally not in temperate countries. We are familiar with bananas, papaya, lychees, guava, citrus fruits, apples, grapes, pears, peaches, etc. However, it is thrilling to sample mangosteens, chompoo, salak, cherimoya, jackfruit, langsat, star apples, breadfruit, pomelo, and noni. I didn't have the courage to try durian, which has a sweet taste but a horrible smell.

For the graduation ceremony of the school Princess Sirindhorn attended. She is the daughter of the late King Bhumibal and sister of the current King. She is fluent in Thai, French, English, and Mandarin. She asked Michele for books on astronomy, which Michele offered to send. I talked with the Princess. She has broad interests. In preparation for the Princess' visit, 200 Thai soldiers were stationed in the woods around the observatory.

37

THAILAND, LAOS, MALAYSIA (2004)

Dr. Chayan Boonyarak, an astronomer at Naresuan University in Phitsanulok in central Thailand between Bangkok and Chiang Mai, obtained funds for a 6-month sabbatical in the US in 2003. That was in connection with his university obtaining a 1 m telescope. Kam-Ching Leung (Univ. Nebraska) recommended that he go to KPNO and I agreed to take him on. We did various projects, of which three were published: (1) one on citations to IAU papers, (2) a study of rotational velocities in binaries was published in the ApJ, and (3) a study of citations to papers 50 years after publication.

When he returned to Thailand he invited me to lecture at his university and at other universities in Thailand. I agreed to go there for five weeks in Jan.-Feb. of 2004. I prepared 14 lectures and added more once I got there. They provided the costs of transportation, hotels, food, and $30-50 per lecture. It was not enough to bring John with me. I started to fly there on 6 Jan., going through LA and Osaka to Bangkok, taking 25 hours.

In Thailand there are five universities that cooperate in curricula, requirements, and planning. A group of professors from Naresuan University planned a 1-week tour with faculty from the other universities. Therefore Chayan met me in Bangkok and we flew to Ubon Ratchathani in eastern Thailand near Laos and the site of Ubon University. I gave a talk there on "Planets around Other Stars". It is on a beautiful lake with large houses along the shoreline, including the vacation house of the King and Queen.

The next day we boarded a bus into Laos, through Pakse (population 120,000, the third largest city in Laos), across the Mekong River in a ferry, to the ancient (1000 years old) temple of Wat Phu. It is 10 km. wide ruin of the Hindu Khymer Empire that produced Angkar Wat in southwest in Cambodia.

After returning to Phitsanulok and Naresuan Univ., the group of faculty members flew (65 min.) on an A3100 to Hat Yai, the fourth largest city in Thailand with a population of 150,000. It is in southern Thailand, about 150 km north of Malaysia. It was in the news years later because the people are mostly Muslim in the predominantly Buddhist country and riots occurred. We stayed at Haad Kaew resort outside of Songkhla. Songkhla is on the east coast of the long Malaysian peninsula of Thailand; it is between a large lake and the sea. We took a ferry across the lake and a tram up Tang Kuan Hill. The next day the 40 faculty members from the five universities met at Hatyai Univ., a private school founded in 1997 and having 5,000 students. I walked along the beach, swam in the resort pool, and read. But I learned that my visa into Thailand was for a single entry only. A lady at Smile Tours took me to the Police Station where they took 1.25 hour and $28 to change my visa to multiple entries.

The next day we crossed the border into Malaysia and visited the new location of Universiti Utara Malaysia (Univ. of Northern Malaysia), which specializes in management. We had a Chinese lunch in Alor Setar (300,000 people) and then across a 13 km bridge to Butterworth (population 72,000). The next day we (the 40 faculty members plus 10 from SUM) met at the Science University of Malaysia (25,000 students) in George Town (220,000 people). I gave a short talk recommending the teaching of astronomy to all science students.

That was the end of our week-long tour of Thai universities.

Chayan and I flew back to Bangkok on a 737. I used the Internet Café (remember those?), costing $0.17 per minute to reach John. Next day we went to the Mahidol University where I met David Ruffolo, who had obtained a Ph. D. at the University of Chicago under Peter Meyer. That is one of the two best (out of dozens) of the major universities in Bangkok, the other one being Chulalongkorn Univ. I gave a talk at Mahidol Univ.

Then to Naresuan Univ. (16 km south of Phitsonulok), named for King Naresuan the Great in 1990 and has 25,000 students. It has very modern buildings. Chayan became the Dean of the Science Faculty while I was there. I taught an undergraduate class of 22 students on stellar astrophysics, five classes per week. My technique for talks in Asia was to talk somewhat slowly, using simple English, and projecting on a screen a PowerPoint version of my talk. In Naresuan Univ. I distributed after each talk a copy of my talk. I found that I can explain stellar interiors and stellar evolution in understandable language. I stayed in the Topland Hotel ($20 per night) on the 15th floor. Attached to it was a 5-story shopping center (with restaurants) and an Internet Café.

One week-end I went to Chiang Mai where Boonraksar "Boon" Soonthorthum, educated at the Univ. Texas, is Chair of the Astronomy Dept. in Chiang Mai Univ. He took me to the new National Astronomical Research Institute (NARI, later named NARIT) on the highest mountain (Doi Inthanon with a labelled height of 2565.3341 m!) in Thailand and 15 km south of Chiang Mai. Below the top is the 1383 AD temple Wat Phra That Doi Suthep. They were building a 2.4 m Ritchey-Chretian alt-azimuth telescope that Princess Sarindhorn later dedicated. She is the daughter of the late King Bhumibol Adulyadej and brother of the current King. She travels much in Thailand and officiates at many ceremonies, like handing out all of the university degrees. Also on the mountain top next to the observatory is the Thailand Air Force Radar Installation. At the Wat Phra That Doi Suthep temple the Air Force had built two new temples in honor of the King (Bhumibol) and Queen.

Then Boon drove me down a steep hill west of Chiang Mai to a Hmong Tribal village. The Hmong Tribe is one of 13 ethnic tribal groups living along both sides of the Thailand-Myanmar border. They migrated many years ago from China to Laos, Vietnam, and Thailand. They have their own language, religion, dress, and customs. They are not citizens of Thailand and have an ambiguous status. They grow poppies for heroin to make money.

If you think that I had made trips to many countries, those were only a drop in the bucket. Here are some of the trips that I have not mentioned: Alaska (1 trip), Argentina (1), Australia (1), Belgium & Strasburg (~5 trips), Cambodia (Angkor Wat), Canada (~4), Chile (7 to Cerra Tololo), China (11 more trips), Columbia (1), England (1), Greece (1), Hawaii (~8, including hiking the Napali Trail and sailing trips with the Boesgaards), Hong Kong (~5), Indonesia (3, including 2 to Bali), Italy (2), Mexico (~5, not including Rocky Point), Netherlands (1), New Zealand (1), Paris (~6, my favorite city), South Korea (~5), and the Vatican City (1). That's ~70 trips besides the ones reviewed. I never got sick on any of my trips except during one hour of Montezuma's revenge in Chile, and my broken leg in Vienna, Austria.

38

CULTURE

When I become interested in a field, I tend to gravitate to the very best in the field. For instance, when I became interested in hand-carved objects, I was attracted to Chinese jade carvings that involved both some of the most beautiful material that is very difficult to carve (see the Section 27 on China).

I have read large sections of the best literature, including Shakespeare, but rarely "best sellers" or popular books. I average reading 20-50 books a year, plus many magazines including *The Economist, Science, Physics Today*, and *Archeology*. My favorite books are *The Bridge of San Luis Rey* by Thorton Wilder and *Winesburg, Ohio* by Sherwood Anderson. Among recent books I recommend *AKA Shakespeare* by Peter Sturrock, in which several scientists and well-read people discuss the evidence for various people as the author of the Shakespeare plays and sonnets. I am currently reading *The Jungle Book* because composer Charles Koechlin wrote many pieces based on it. I learned that the *The Bander-log* was the tribe of monkeys that kidnapped the man-cub until his friends Baloo, the brown bear, and Bagheera, the black panther, rescued him. Delightful!

In travel I gravitated to Asian countries and exotic places such as Easter Island, Canyonlands, Petra, Micronesia, the Galapagos, and Fiji, avoiding well-known places like Tal Mahal, famous waterfalls, or standard cruise trips.

In music too I go for the best. In symphonies I like best those of Mahler, Brahms, Shostakovich, and much contemporary classical music such as that of George Crumb and John Cage. I remember that even in high school my friends and I were thrilled with the new Shostakovich symphonies as they came out. In piano music I love that of Liszt, Brahms and Ravel, the sonatas of Beethoven and Prokofieff, and the *Preludes and Fugues* of Shostakovich. In quartets the best seem those of Beethoven, Bartok, Shostakovich, plus many others. I have been on the Board of Directors of the Arizona Friends of Chamber Music for 30+ years. We bring to Tucson the best musical groups in the world for ~20 concerts a year and have commissioned more than 60 new chamber music pieces. I grew to like chamber music from the Sunday evening concerts in Dabney Lounge at Caltech. In songs I like those of Schubert, Mahler, Ravel, and Debussy. I am bowled over by the operas of Wagner and the four operas of Orff, but my favorites are *The Tales of Hoffmann, Pelleas et Melisande, Tristan and Isolde, Das Rheingold,* and *Wozzeck*. Popular music, rock, and country music bore me by the triteness of their texts ("I love you, you love me, dah, dah, dah") and monotony of the music (although dancing requires a repetitious rhythm). Classical music is much more interesting with varied rhythms, larger variations in volume from barely audible to extremely loud, greater variety in instrumental sounds, a wide range of moods, etc. Styles change. Whereas in the 20th century the emphasis was on themes and modifications of them, in the 21st century the emphasis is on interesting sounds. I like thoughtful and expressive music.

The one piece of music that has been my favorite for at least 60 years is Gustav Mahler's *Das Lied von der Erde* (*The Song of the Earth*), based on the Tang Dynasty poetry of Li Bai (mostly). I find that Mahler's music is the most profound and inspiring music of all composers.

An exception to liking the best is in food. In drinks I did not find beer or wine as refreshing as cokes. Also, as described in the Health section, my weird tastes in food meant that I do not relish interesting dishes. The Chinese have an extreme variety in their foods because they are so inventive. Initially in the feasts that were given for me as a distinguished foreigner, they served exotic foods. My notes for one dinner said that they served grasshoppers, fried mushrooms, deer tendons, deer meat, pigeons, shell fish, *Pass Over Bridge Rice Noodles*, moon cakes, etc. I will admit that I enjoyed shark fin soup and bird nest soup, but after I learned how they were made, I avoided them on ethical grounds. Later the large dinners in my honor included more normal Chinese dishes, usually 18 or more in number.

I am bored with sports. Who cares which team carries the ball to the end of the field? What difference will it make to the world the next day or month? All this meant that I did not share the interests of most people (sports, beer, rock, religion, cars, etc.). That is why I had difficulty making friends, although I do not judge others for what they like or do. I tend to like virtually all people I met and do not like to argue with others. I liked Corvette convertibles because of their beautiful shapes and had three (1968, 1972, 1976) until my mother needed wheel chairs, which are hard to put on a Corvette. I also liked my sailboat on Lake Geneva.

39

IN MY HOUSE

My house contains some of my favorite things, obtained from all over the world. Let me tell you about some of them.

I always liked the Peanuts cartoons of Schulz. They have the kind of humor that appeals to me: the ridiculous things that happen to people in everyday life. Some of the cartoons involved astronomical themes. My friends knew of my liking for Peanuts. One person – H. John Woods – knew and shared that sense of humor (see the story about him in the Addenda). For reasons that I do not remember, he felt obligated to me for what I had done for him. He wrote to Charles Schulz in California and told him about his astronomical friend who liked his cartoons, particularly the astronomical ones, and asked for an original strip.

Schulz replied that he didn't have any astronomical ones and most of the original strips were at the United Features Syndicate, Inc. So he sent John's letter to the Syndicate. The Syndicate wrote to John saying that they didn't have any astronomical ones, but if they knew Schulz, he would draw one soon. Some weeks later one appeared in the newspapers, completely out of sequence, of Snoopy sitting on his doghouse in the first three frames and looking at the sky in various directions. In the fourth frame Snoopy thinks 'I wonder if there are "people" stars and "dog" stars?' Shortly thereafter the Syndicate sent John the original, with a greeting from Schulz to me written on the side of the doghouse, and John sent it to me. I didn't know of any of that correspondence, but the original 1965 strip and the copy in the newspaper are framed in my bedroom.

I cannot read Chinese calligraphy; one has to be younger than 10 years to start to memorize the ~3000 characters that a typical Chinese knows. However I can appreciate the different styles used. Once when I was in the gift shop of the Forest of Stele in Xi'an, I saw a vertical scroll with a regular style that I immediately liked. The artist was behind the counter. I congratulated him and bought it for the $100 asking price. A Chinese friend later told me what it said. It was one of Li Bai's best known poems from the 8th century called *Invitation to Wine*. It starts (in translation):

> "Do you not see the Yellow River come from the sky,
>
> Rushing into the sea and ne'er come back?
>
> Do you not see the mirrors bright in chambers high,
>
> Grieve o'er your snow-white hair though once it was silk black?
>
> When hopes are won, oh! drink your fill in high delight,
>
> And never leave your wine-cup empty in moonlight!
>
> Etc."

Li Bai was an alcoholic and wrote some of his best poems while he was drunk. He also wrote most of the six poems used by Gustav Mahler in his *Das Lied von der Erde*. I enjoy looking at that calligraphy.

Another favorite Chinese scroll is in my office. It is a poem written in the 8th century by Zhang Ji called *Mooring at Night near the Maple*

Bridge. The scene is in the canal city of Suzhou. I bought it in Xi'an in 2017. I am told that all Chinese students learn to recite that poem by heart.

I enjoy Chinese opera. It is stylized with extremely colorful costumes, acrobatics, weird music, and an English translation on a board above the stage. I can just follow the plots. Actually, I have become very fond of the voice of the best-known opera singer, Mei Lanfang (1894-1961), and have videos of all of his opera roles. Usually when in Beijing my friend Xiang-Tao He takes me to a performance. He is an astronomy Professor at Beijing Normal University, where I give talks to his classes. He is extremely knowledgeable about Chinese opera. His recent astronomical work has been on quasars and he spent some time in Tucson, working with Richard Green.

Once when we went to the opera, there were some jade carvings available for sale in the lobby. One attracted me. It was a single rock seven inches (18 cm) high with colors of black, white, and light green. The black in back represents a background rock. The white in front shows a tree with intricate branches and flowers. The light green in the upper right shows two birds in the tree. Beautifully carved! Xiang-Tao helped me buy it. One of my best jade carvings.

In Tokyo I bought an original signed woodblock print by Kiyoshi Saito of a village scene under heavy snow on Hokkaido, the northern island of Japan.

I always liked sculpture, both realistic (Greek) and abstract (e.g. Henry Moore, Barbara Hepworth, Brancusi, Arp). I bought a bronze copy (20 inches = 50 cm) high of a Greek stature found in Marathon Bay in 1925. The original is about five feet (1.5 m) high and is called "The Marathon Boy". There is no written record of it but the style places from about 330 BC. It is a statute of an *ephebe*. He is holding one hand out, palm up, perhaps feeding birds out of his hand. The other arm is held up, palm out, perhaps cautioning others to stay away. His head has thick hair held in place with a head band.

Easter Island has the only written language among the Pacific islands. It is a picture language written on wooden boards about a half-meter long called rangorango boards but has never been deciphered for two reasons. One is that in the 1870s many of the men on the island were taken as slaves to work in bat guana caves off the coast of Peru. Many died there and when, due to international outbursts, the remainder were returned to Easter Island, they brought diseases that killed all but 123 islanders. So few people were left who were able to read the rangorango boards. The second reason is that Christian missionaries convinced the islanders that the rangorango boards were heathen things and should be destroyed. As a result only about 15 originals remain, too few for scholars to decipher. On Easter Island I traded my tape recorder for a copy of a rangorango board. Fascinating script!

In Paris there is a company called Baccarat that makes fine crystal glass. I have bought three pieces, all designed by the same man. The one in my bedroom is a heavy (5.5 lbs = 2.5 kg) leaded-glass vase six inches high and two inches thick (15x5 cm). It is pear-shaped with a flat front surface and one inch squares incised on the back. The two in my office are called La Soliel (the Sun) and Windsail (like the sail on a sailboat). The leaded glass really sets off the metal detectors at the airport!

In the nightstand next to my bed is a notebook in which I record the author, titles, and dates of the books I have read. It provides a way of looking up books I particularly liked. Since 1948 I have averaged reading 23 books per year, although sometimes I read long or short books. The reading is usually between when I come home at ~11 pm and when I go to sleep at ~1 am. There were fewer books when I was an editor or observing or traveling and when John recorded TV shows for us to watch when I came home. My library shelves at home have ~2,000 books.

Also in my house are an original lithograph by Juan Miro, Easter Island statues, a painting on sandstone from Canyon de Chelly, a painted tea cup with cover from China, pictures of John, six Paul Klee paintings, a picture of Wright's *Falling Water*, Rousseau's *Sleeping Gypsy*, John Rogers Cox *Gray and Gold*, and many other things.

40

HEALTH

My health has always been good after the usual childhood diseases. I had whopping cough at age 4 or 5 that left me with a strange appetite. After that I never liked fat or greasy or rich foods, like butter or salad dressings. Only plain foods. Also later I never was tempted to smoke because it seemed unnatural and stupid to inhale smoke into one's lungs. I tried wines and beers, but didn't like them and, again, thought that it cannot be good to pour harsh chemicals into ones system, particularly when I saw people becoming addicted to them. After moving to the Southwest I got plenty of vitamin D from sunshine, so I never had a cold or the flu (even without vaccinations) or any other sicknesses after the age of 35.

There is another reason why I do not get sick. Most people avoid catching diseases from other people by not touching surfaces (doorknobs, push plates, railings, etc.), and never putting their fingers into their mouths. I touch all surfaces and often stick my fingers into my mouth. As a result, I am often acquiring germs and challenging my immune system to be strong.

After buying the condo in Orchard River, I went swimming in my Speedos, doing laps, every morning from April to October. Heart disease ran in my family of the male side. My father died of a heart attack at 50 and his father did the same. Both my brother and I had bypass surgery and I continued to take three prescriptions, but never had further problems. I never had arthritis, back, or joint pains or any mental problems such as depressions. What a lucky person!

However, I have broken bones and other operations. In 1960 at an AAS meeting in Mexico City, I slipped on a comb lying on a cobblestone sidewalk and broke my right arm, just before I was to give the first paper of the conference. Fortunately I had written out the complete text and Harlon Smith read it for me. In 1966 in Pasadena another driver ran a red light at a blind corner and hit the side of my car. I received a cracked pelvis, two broken ribs, and a broken collar bone, which still has a screw in it. In 1996 I had a double by-pass surgery and in 2008 a triple by-pass.

In 2002 I broke my right foot badly while falling with a ladder while replacing lights. Dr. Bradley Brainard repaired it and left 10 screws in it. He said that it would probably never be normal again. But about five years ago I wrote him to say that I couldn't remember which foot he had worked on because both of them were the same. He replied that I had to take part of the credit for the recovery because I did not favor that foot.

In 2018 I broke my left leg at a hotel in Vienna, Austria where I went to attend an IAU meeting. I never made it to the meeting. They inserted a steel rod in my femur in a way that they would not do in the US because patients would freak out. They injected an anesthetic into my spine that made me numb below the waist but I was awake during the operation and was able to talk with the doctors (in English). I felt them cut into my leg and heard them pound the steel rod into my femur with a steel hammer. But it took me a long time to recover normal walking ability because I spent too many months lying on beds in hospitals and rehabs, so my muscles tightened up. At latest count, I have at least 19 screws and metal objects in my body (bionomic man?).

This has been a difficult time because of the pandemic when we do not know which of the people we encounter can spread the disease. But I remember a similar situation in the 1930s when thousands of kids and young adults contacted polio and we did not know how it spread.

Some of those people, such as Arthur Clarke, had serious muscle pains 40-50 years later in the muscles that substituted for paralyzed ones and had worn out.

I am not religious. In fact, when I see the wars, killings, and hatreds caused in the name of religions, I wish that they would all disappear. Also, they do not make sense to me, e.g. after Noah's flood, how did the kangaroos get back to Australia? And how did Noah put in his small boat pairs of the 100,000 different species of animals? Mark Twain said that "Faith is believing something that you know is not true".

The attraction of religions is the promise of life after death. But for how long? They say for eternity. How long is that? Hundreds of years, thousands, millions, billions? It would be great to see lost friends and relatives again, but what does one do after a couple hundred years and you have mastered all of the harp literature? I have many interests, but in this life I doubt that I would find new interests to last me 200 or 300 years. After that I would be bored stiff. To me, living thousands or millions of years would be worse than death. I prefer to die dead than to live for an eternity. For me the Golden Rule is an adequate guide for moral behavior.

41

SUMMARY

I have been lucky enough to live at a time when astrophysics was developing to the point of understanding the basic evolution of stars. My part was to look at some of the abnormal stars (Ap, Am) to see how they fit in and to help understand the role of duplicities. The latter happened to lead to the discovery and study of exoplanets. In a way, it is unfortunate that I did not concentrate on a single field, become a world's expert, and win important prizes. After finishing one project, I was tempted to do something different, but would come back to fields later. I always had enough imagination to foresee breakthroughs and start new fields. Those include studying the role of the frequencies of binaries, what we can learn from publication studies, the content of open clusters, the differences between Population I and II stars, etc. In fact, I like to do research so much that I neglected reading about what others have done.

Also this was a time when I knew many of the astronomers and could enjoy them as friends, partly in my role as Editor of the world's largest and most important astronomical publication. In that role I helped make the process of publication faster and more fair to the authors, referees, and publishers. I liked the roles of doing research, explaining it to others, and helping to publish the results. I also liked the benefit of worldwide travel. I made 80 or more trips abroad, visiting about 30 different countries, and made many friends abroad. The

14 trips to China were especially interesting because of my chance to learn about their rich culture that has lasted more than 5000 years. It is a shame that the history that we learn in schools is almost all about Europe and the Americas, when the Chinese culture has exceeded and pre-dated ours in so many areas. I cannot understand the arrogance of educators who think that we are the best in all ways, although we have the one advantage of allowing and encouraging free individual thought, so that the western countries still exceed in new ideas.

I always had wide interests. Science is great, but so is worldwide literature, classical music, art, poetry, travel, metaphysics, psychology, physics, and fields about which I know very little like biology and chemistry.

I tend to like all people and do not care to engage in arguments. For instance, at Yerkes I was Assistant (Associate? I do not remember) Director because the primary user of the shops could not get along with the workers so I acted as intermediary because I can get along with everybody. It helps to have a sense of humor.

REFERENCES

Aarseth, S. L., Hills, J. G. 1972, A&A, 21, 255

Abt, H. A. 1955, ApJ, 122, 72

Abt, H. A. 1957, ApJ, 126, 138

Abt, H. A. 1961, ApJS, 6, 37

Abt, H. A. 1965, ApJS, 11, 429

Abt, H. A. 1970, ApJS, 19, 387

Abt, H. A. 1973, ApJS, 26, 365

Abt, H. A. 1978, in *Protostars & Planets*, ed. T. Gehrels (Tucson: Univ. Arizona Press), 323

Abt, H. A. 1979, 320, 485

Abt, H. A. 1980, PASP, 92, 249

Abt, H. A. 1981a, ApJS, 45, 437

Abt, H. A. 1981b, PASP, 93, 269

Abt, H. A. 1982, PASP, 94, 213

Abt, H. A. 1983a, PASP, 95, 113

Abt, H. A. 1983b, Ann. Rev. Astron. Ap, 21, 343

Abt, H. A. 1984, PASP, 96, 746

Abt, H. A. 1985a, ApJ, 294, L103

Abt, H. A. 1985b, ApJS, 59, 95

Abt, H. A. 1987a, PASP, 99, 439

Abt, H. A. 1987b, ASP Leaflets, 1329

Abt, H. A. 1988a, PASP, 100, 506

Abt, H. A. 1988b, PASP, 100, 1567

Abt, H. A. 1990a, PASP, 102, 368

Abt, H. A. 1990b, PASP, 102, 1161

Abt, H. A. 1990c, ApJ, 357, 1

Abt, H. A. 1991, *Science*, 251, 1408

Abt, H. A. 1992a, PASP, 104, 235

Abt, H. A. 1993, PASP, 105, 437

Abt, H. A. 1995a, PASP, 107, 401

Abt, H. A. 1995b, ApJ, 455, 407

Abt, H. A. 1996, PASP, 108, 1059

Abt, H. A. 1998, *Nature*, 395, 756,

Abt, H. A., Editor, 1999a, ApJ 438, 1

Abt, H. A. 1999b, AAS, 195,1300

Abt, H. A. 2000a, PASP, 112, 1417

Abt, H. A. 2000b, BAAS, 32, 937

Abt, H. A. 2001, AAS, 199, 1450

Abt, H. A. 2002, BAAS, 34, 1354

Abt, H. A. 2005a, BAAS, 37, 551

Abt, H. A. 2005b, BAAS, 37, 1540

Abt, H. A. 2006, AAS, 335, 169

Abt, H. A. 2007a, *Scientometrics*

Abt, H. A., 2007b, *Scientometrics*

Abt, H. A. 2007c, AAS, 211, 7105

Abt, H. A. 2008a, ApJS, 176, 216

Abt, H. A. 2008b, AJ, 135, 722

Abt, H. A. 2009a, ApJS, 180, 117

Abt, H. A. 2009b, AJ, 138, 28

Abt, H. A. 2009c, PASP, 121, 1291

Abt, H. A., 2009d, PASP, 121, 544

Abt, H. A. 2010, PASP, 122, 1015

Abt, H. A. 2011a, AJ, 141, 165

Abt, H. A. 2011b, BAAS, 24, 77

Abt, H. A., 2012, AJ, 144, 91

Abt, H. A. 2014, PASP, 126, 409

Abt, H. A. 2015, PASP, 127, 713

Abt, H. A., 2016, PASP, 128, 4501

Abt, H. A., 2017a, PASP, 129, 4008

Abt, H. A. 2017b, PASP, 129, 4505

Abt, H. A., 2018, PASP, 130, 4506

Abt, H. A. 2019, BAAS, 51, 0207

Abt, H. A., Biggs, E. S. 1972, *Bibliography of Stellar Radial Velocities* (Kitt Peak National Observatory)

Abt, H. A., Boonyarak, C. 2004, BAAS, 35, 948

Abt, H. A., Fountain, J. W. 2018, Research in Astron. & Ap, 18, 37

Abt, H. A., Garfield, E. 1992, *Scientometrics*

Abt, H. A., Garfield, E, 2002, *Scientometrics*

Abt, H. A., Jeffers, H. M., Gibson, J., Sandage, A. R. 1062, ApJ, 135, 429

Abt, H. A., Levy, S. G. 1976, ApJS, 30, 273

Abt, H. A., Levy, S. G. 1978, ApJS, 36, 241

Abt, H. A., Meinel, A. B., Morgan, W. W., Tapscott, J. W., 1968) Chicago: Yerkes Obs.)

Abt, H. A., Morgan, W. W., Strömgren, B. 1957, ApJ, 126, 322

Abt, H. A., Zhou, H. 1996, PASP, 108, 375

Abt, K. W. 2004, *A Few Who Made A Difference* (NY: Vantage Press)

Aitkin R. G. 1932, *New General Catalogue of Double Stars* (Washington: Carnegie Inst. of Wash., Publ. 417)

Babcock, H. W. 1958, ApJ, 128, 228

Boss, A. P. 2003, in IAU Symp. 211: Brown Dwarfs, 23

Branch, D. 1976, ApJ, 210, 392

Brandt, J. C., Williamson, R. A. 1979, J. History of Astron. Suppl., 10, S1

Campbell, B., Walker, G. A. H., Yang, S. 1988. ApJ, 331, 902

Clark, D. H., Stephenson, F. R. 1977, *The Historical Supernovae* (Oxford, New York: Pergamon Press)

Collins, II, G. W.,, Claspy, W. P., Martin, J. C. 1999, PASP, 111, 871

Deutsch, A. N., Lavdovsky, V. V. 1940, Poukobo Obs. Cir, 30, 21

Duncan, J. C. 1921, Proc. Nat. Acad. Sci., 7, 179

Duyvendak, J. J. L. 1942, PASP, 54, 91

Duquennoy, A., Mayor, M. 1991, A&A, 248, 485

Fountain, J. 2000, *Rock Art Papers*, San Diego Museum Papers, 39, 91

Garmany, C. D., Conti, P. S., Massey, P. 1980, ApJ, 242, 1063

Gliese, W. 1969, Veröff, Astron. Rechen Inst. Heidelberg, no. 22

Greenstein, J. L. 1948, ApJ, 107, 51

Gum, C. S. 1952, Observatory, 72, 151

Halbwacks, J.-L., Mayor, M., Udry, S., Arenou, F. 2004, Rev. Mex. A. A., Ser. de Conf., 21, 20

Hoffleit, D., Jaschek, C. 1982 *The Bright Star Catalogue, 4th ed. revised* (New Haven: Yale Univ. Obs.)

Hoffleit, D., Saladyga, M., Wlasuk, P. 1983 *A Supplement to the Bright Star Catalogue* (New Haven, CT: Yale Univ.Obs.)

Hubbard, E. N., Dearborn, D. S. P. 1980, ApJ, 239, 248

Joy, A. H., Abt, H. A. 1974, ApJS, 28, 1

Kovtyukh, V. V., Wallerstein, G., Andrievsky, S. M., Gillet, D., Fokin, A. B., Templeton, M., Henden, A. A. 2011, A&A, 526, 116

Kroupa. P. 1995, MNRAS, 277, 1491

Lampland, C. O. 1921, PASP, 33, 79

Latham, D. W., Stefanik, R. P., Torres, G., David, R. J., Mazeh, T., Carney, B. W., Laird, J. B., Morse, J. A., 2002, AJ, 124, 1144

Lundmark, K. 1921, PASP, 33, 225

Mayall, N. U., Oort, J. H. 1942, PASP, 54, 95

Mayer, D. 1979, Archeoastronomy, 1, S53

Mayor, M., Queloz, D. 1995, *Nature*, 378, 355

McCrea, W. H. 1964, MNRAS, 128 147

Meinel, A. B. 1958, Contr. Kitt Peak Nat. Obs., No. 78

Mermilliod, J.-C. 1982, A&A, 109, 37

Michaud, G. J. 1970, ApJ, 160, 641

Miller, W. C. 1955, ASP Leaflets, 7, 105

Morgan, W. W., Abt, H. A., Tapscott, J. W. 1978 *Revised MK Spectral Atlas for Stars Earlier then the Sun* (Chicago: Yerkes Obs, Univ. Chicago & Kitt Peak National Obs.)

Morgan, W. W., Keenan, P. C., Kellman, E. 1943, *An Atlas of Stellar Spectra*, (Chicago: Univ. Chicago Press)

Pendl, E. S., Seggewiss, W. 1975, in *IAU Colloquium 32, Physics of Ap Stars*, ed. W. W. Weiss, H. Jenkner, H. J. Wood (Vienna: Univ. Vienna), p. 357

Phillips, J. A., Clegg, A. W. 1992, Nature, 360, 137

Porter, E. 1968, *Galapagos, The Flow of Wildness*, Vol. 1 and 2 (A Sierra Club-Ballantine Book)

Pourbaix, D. et al. (nine authors), 2004, A&A, 424, 727

Preston, G. W., Sneden, C. 2000, AJ, 120, 1014

Sanford, R. F. 1949, PASP, 61, 135

Schüssler, M., Pähler, A. 1978, A&A, 68, 57

Slettebak, A. 1954, ApJ, 119, 146

Slettebak, A. 1970, in *IAU Colloquium 4, Stellar Rotation*, ed. A. Slettebak (Dordrecht: Reidel), p. 3

Smith, R. K., Bautz, M. W. et al. 2007, PASJ, 59, 141 (21 authors)

Tassoul, J.-L, 1978, *Theory of Rotating Stars* (Princeton: Princeton Univ. Press)

Tremaine, S., Dong, S. 2012, AJ, 143, 94

Trimble, V. 1968, AJ, 73, 535

Walsh, B. Y., Lallement, R. 2008, A&A, 490, 707

Wilson, O. C., Abt, H. A. 1954, ApJS, 1, 1

Wilson, R. E. 1953, *General Catalogue of Stellar Radial Velocities* (Washington: Carnegie Inst. of Wash. Publ. 601)

ADDENDA
STORIES ABOUT ASTRONOMERS

COLLECTED BY HELMUT A. ABT

INTRODUCTION

Astronomers have excellent senses of humor. In the past, theirs was a small community where virtually each one knew all the others personally. Unfortunately, most of the astronomers in these stories have left us; for the remainder I got their permission to have these stories included. All these stories are true to my knowledge.

They are arranged in alphabetical order, so we can laugh at me first.

HELMUT ABT'S BROKEN ARM

One day George Backus came back to Yerkes Observatory from the University of Chicago campus where he had just passed his Ph.D. final exam. He brought a bottle of wine and insisted that all the dinner guests in Mrs. Van Biesbroeck's rooming house toast him. Helmut Abt, a teetotaler, took a polite sip. After dinner while walking to the Observatory, Helmut slipped on the ice and broke his arm. No one would believe that he wasn't stinking drunk.

THE SKUNK SMELL ALL NIGHT

One night Helmut and a night assistant worked in the open 2.1m dome. They smelled skunk all 12 hours of the night. It was awful! Finally at dawn Helmut discovered the problem. Some workmen had been building a low wall near the dormitory and dropped their lunch remains in a steel can next to the wall. A skunk jumped from the wall into the can, and was trapped all night. He used his only defense mechanism. Helmut took a long 2x4 and knocked over the can. The skunk walked out, glaring at Helmut as though he caused the whole thing.

PAYMENT TO REFEREES

Helmut gave a talk to a society of the editors of the 60 scientific journals produced in Shanghai. In the question session he was asked how much do American journals pay their referees? He said, "Nothing. They expect the referees to help other authors improve their papers in exchange for others having helped the referees improve theirs. We would consider paying referees too capitalistic a system for us." To which the Chinese replied, "We find your system to be too communistic for us."

OIL CAN ON THE 0.9M TELESCOPE MIRROR

Helmut was scheduled to use the 0.9-meter telescope and spectrograph. Because it was the monthly visitors' night on the mountain, the first hour or two was to be spent showing objects to visitors. He looked through the eyepiece and the images were terrible and couldn't be focused. After some searching, he climbed the ladder on the inside of the dome and looked down at the mirror. There was an oil can on the mirror, and as the telescope was moved, the oil can spread oil over the surface of the mirror.

What had happened was that a week earlier a mechanic had climbed up that ladder to oil the shutter mechanism and unintentionally left the oil can up there. In subsequent use, the dome motion shook the oil can off the ladder and down the telescope tube. Fortunately it did not hit the mirror, but just slide down the side of the tube and gradually came to rest on the mirror. But what bothered Helmut was that observers had used the telescope for a week before he came on and did photometry without noticing the bad images.

Helmut's Cokes

Helmut kept astronomer's hours, waking up at noon and going to bed past midnight. Working alone in his office, he noticed severe irregularity in his pulse. So off to the local doctor for a visit. The doctor told him to quit drinking coffee. "I never drink coffee," said Helmut. Well then stop drinking so much tea. "I never drink tea," said Helmut. So the doctor told him to quit smoking, to which Helmut said "I don't smoke, and never have." Well then, the doctor said, give up drinking alcohol. Helmut replied "I never touch alcohol, and never have." Well then, the doctor said, "What do you do"? Well, I drink a lot of Coca Cola. Aha, the doctor said, that must be it, too much caffeine! So Helmut cut back on the coke, and his pulse became regular again. Being a scientist, Helmut returned to drinking coke, just to confirm the experiment, and sure enough, the irregularity returned, and so Helmut cut back permanently.

HELMUT'S KIDNEY STONES

A similar story is that in the 1960s Helmut started having trouble with kidney stones. They were painful. But what caused them? Helmut noticed that they occurred only in the spring each year. But what was he doing in the spring? Well, he went to San Francisco for ASP Board meetings. Being a chocoholic, he went to Ghirardelli Square and bought several pounds of milk chocolate, which he ate too quickly when he returned to Tucson. Was that causing the kidney stones?

So he stopped eating Ghirardelli chocolate and there were no more kidney stones. But was that just a coincidence? So he bought some Ghirardelli chocolate and the kidney stones returned. Q.E.D.

WALTER ADAMS IN THE WASHROOM

Walter Adams was Director for many years of the Mt. Wilson Observatory, a part of the Carnegie Institution of Washington. He prided himself on staying within his budgets and even returned small amounts to Carnegie. Once he attended a meeting in Washington of the directors of the various Carnegie Institutions. In the men's room he mentioned to another man that the men's room was very luxurious with marble walls and expensive brass fixtures. The other man, not knowing Adams, said, "You won't believe this, but there is an old director in California who returns some of his budget every year, so that is why they can afford this."

NOEL ARGUE'S HIDDEN ROOM

Noel Argue lived in an old 16th century house in England. By measuring the rooms within the house and the outside dimensions, he knew that there was a hidden room in the house with no access to it. On cold winter nights around the fireplace, he and his wife would speculate about what might be in that room. Did it have dead bodies or a secret treasure? It was not until some years later that they cut through the exterior walls and found that the hidden room was – empty.

HALTEN ARP'S REACTION TO PICTURE LOCATIONS

In the middle 20th century the only way to put halftones on glossy stock within a paper in the Astrophysical Journal was for someone to hand cut each signature and insert the halftones. That was expensive. So in the 1960s Chandrasekhar started the policy of putting all the halftones on glossy stock at the end of the issue. To Halten (Chip) Arp the halftones were the most important parts of many papers, so he said to me, "Why don't they put the equations at the end of the issue?"

THE MOVIE BOOK

In 1932 Robert Bacher (later head of the Physics Dept. at Caltech) and Samuel Goudsmit (later Editor of the Physical Review) published a book called "Atomic Energy States", giving the nuclear energy levels for the known atomic levels in dozens of atoms. It was a detailed reference book and did not sell many copies. Bob Bacher told me that he and Sam called it their "movie book" because the royalties from the sales occasionally allowed them to go to the movies.

BILL BAUM AND THE NUN

Walter Baade had a nun working for him, getting light curves of variables in M31. One day Bill Baum was puzzling over a photograph he took. He decided to take it down to the measuring room to make some measurements. He stared at the plate as he walked down the corridor, down the stairs, and over to the measuring room. He automatically reached out to pull off the black cloth that covers each measuring engine. But just as he had his hand over the nun's head, he realized his mistake and dashed back up to his office. It took him 20 minutes for his red face to return to its normal color.

THE MEMBER OF THE ROYAL ASTRONOMICAL SOCIETY

Chris Bjornarud was a graduate student at Caltech when I knew him. Previously, as a Canadian, he had been in the Royal Canadian Air Force early in World War II, stationed in England. Being interested in astronomy, one evening he went to the Greenwich Observatory to see the telescopes. He knocked on the front door, to no response, and then on all the other doors that he could find. Finally an elderly gentleman opened a back door and on hearing of Chris' interest in seeing the telescopes, he gave him a tour. On parting he mentioned that if Chris was free the next day, there was to be a meeting of the Royal Astronomical Society at Burlington House.

Chris went to the meeting the next day and was surprised to see that his tour guide of the night before was the Astronomer Royal, Sir Harold Spencer Jones. During the course of the meeting Sir Harold nominated Chris for membership in the RAS, and, of course, it was approved.

After the war while Chris was an undergraduate student at the University of Washington, the elderly professor Theodor Jacobsen was informed that his lifelong ambition was approved to become a member of the RAS. He was so elated that he told everyone he met, including lowly undergraduate Chris. Chris' reply was, "Oh that. I'm a member of that too."

IVAN BRUNETTI'S CARTOONS

Ivan Brunetti was a copyeditor for the Astrophysical Journal as well as an undergraduate at the University of Chicago. He was a skilled cartoonist and his cartoons in the school newspaper were very popular. The cartoon I remember shows a street corner in a city. On the right is a man holding up a sign saying "Jesus Loves You". Around the corner is Jesus holding up a sign saying, "No, I don't".

THE CANDLESTICKMAKER PAPER

S. Chandrasekhar worked very systematically in a field (e.g. radiative transfer, stellar dynamics, hydromagnetics, turbulence) for about five years, wrote dozens of papers, and then summarized those papers in a book.

In 1956-7 John B. Sykes came from England to work with Chandra on hydromagnetic problems. He noted Chandra's style in his papers and wrote a parody. Normally Chandra was very serious, but in this case he saw the humor in it. He had the University of Chicago Press print it as a reprint and individuals at Yerkes paid for copies. The reprints were so widely distributed that finally it was published in the Quarterly Journal of the Royal Astronomical Society, 13, 63, 1972 with an introduction by John. However, John did not explain all the humorous aspects of it, so I will add more explanation.

Nearly all of the references were to volume 237, page 476. At that time only the Proceedings of the Royal Society had that many volumes, so if you look at that volume and page in the Proc. Roy. Soc., you will see the paper that John parodied.

In the Candelstickmaker paper the received date was when Chandra was born. The credit for computing help to Miss Canna Helpit mimics the name of Chandra's computing assistant Donna Elbert. John claims that all the journal names are real ones. The acknowledgements at the end are nearly identical to the ones in Proc. Roy. Soc., 237, 476, except for the puns.

John had a remarkable gift for languages and loved to spend evenings reading foreign language dictionaries. He could read many languages. He later worked as a translator and was Chair of IAU Commission 5 on Documentation. He collected the astronomical papers published in 1881-98 that were available on microfilm.

ARTHUR CLARKE'S VISIT

The University of Arizona invited Arthur Clarke to come for a showing of the film "2001" based on his book. To a packed audience in the Modern Languages Auditorium, he talked a while before his film was shown. By previous arrangement Michael Snowden and I then took Clarke to Kitt Peak where the 4-meter Mayall telescope was in the final testing stages. They put Clarke in the prime focus cage and showed him various astronomical objects. He was so thrilled with what he saw that he said later that he could not sleep the rest of the night.

Meanwhile, Nick Mayall and Dave Crawford were testing the telescope. The field was so large that Nick could photograph M31 and M32 on one plate. Before Clarke got on the telescope, Nick photographed the two galaxies through a red filter. Then he gave Dave the plate to develop. Dave did so, and came out of the darkroom, looking sheepishly, and reported that the plate showed the Coma cluster with double images. How could the telescope be pointed to the north but the plate showed a cluster near the equator? It took a while to figure that one out. On a previous night Nick took a photo of the Coma cluster on a blue-sensitive plate but while inserting the dark-slide, he realized that he jiggled the camera and that the plate would have double images. The plate holder was laid aside, and it was mistakenly given to Nick to photograph M31 and M32. Because Nick used a red filter, he added no blue light to the plate and it showed the Coma cluster with double images.

So Clarke witnessed the frustrations and problems that astronomers sometimes encounter.

DINNERS AT MCDONALD

Anne Cowley had to leave home for 14 days for an observing run, leaving Charles to take care of the children. The kids persuaded Charles to eat dinners at McDonalds for 13 night straight. Then a neighbor felt sorry for Charles and invited them over for dinner the last night.

ARMIN DEUTSCH'S DEAF NEIGHBORS

Once Armin Deutsch lived in an apartment building. The man living below was rather deaf and played his radio very loud. One day Armin looked out of his living room window and noticed a wire going up the outside of the building. He guessed that it was an antenna wire for the man's radio. So he spliced in a rheostat. When the man turned up the volume too high, Armin turned down the volume.

ARMIN DEUTSCH'S DOG

Armin Deutsch claimed that his dog could do arithmetic. To demonstrate that, we would say, "Phido, sit up." And Phido would sit up. Then he said, "Phido, how much is four……plus six……minus ten?" Phido didn't say anything. So Armin said, "See, he knew the right answer!"

FRANK EDMONDSON'S SLIDES

Frank Edmondson was invited to the Greenstein's for dinner. After dinner he offered to show some of his Kodachrome slides. The first slide was of his wife Margaret standing in his front yard. The second was of her standing in the back yard. The third was her on the porch. And so on, and so on. After 15 minutes of dead silence, Jesse Greenstein felt he had to say something, so he said, "Nice slides, Frank." Frank said, "You like them? I have 500 more in my car."

OLIN EGGEN AND THE SWEDE

While Olin Eggen was working at the Lick Observatory, they had an extended visit from a Swede. Because both of them kept night hours, the Swede got well acquainted with Olin. One day the Swede asked Olin, "Tell me, Olin, is there a word in English that I can use when things don't go right in the dome?" Olin thought a minute, and then said, "Yes, but you should use it only when you are alone in the dome." The Swede agreed, so Olin told him that it was, "Oh, fudge." Later the Swede reported that the word worked well.

One evening both of them were at a dinner party, when Olin spilled his drink. He said, "Oh, fudge!" The Swede got so embarrassed for him that they had to explain what it meant.

THE BOTTLES OF RHINE RIVER WATER

Russell A. Fischer was my thesis advisor for my Master's thesis at Northwestern University in 1948. But during WWII he was a member of Samuel Goudsmit's ALSOS team to determine whether the Germans were working on an atomic bomb. They reasoned that if they were doing so anywhere in western Germany, some radioactive material would get into the Rhine River that drained all of western Germany. So they needed to get samples of the river water near its outlet in the Netherlands. Unfortunately, this needed to be done during the Battle of the Bulge when that region was under attract. So Russell Fischer was told to crawl across a Rhine River bridge, under enemy fire, and to lower bottle into the water. He filled three bottles. Not having a means of testing for radioactivity in the field, they flew the bottles to Washington for analysis. Their box allowed the bottles to rattle around, so they added a bottle of French wine to fill the box and put on a note saying "Test this for radioactivity too."

The report came back that there was no radioactivity in the Rhine River water (i.e. no attempt by the Germans to build an atomic bomb in western Germany), but there was activity in the French wine, so send more samples! They brushed that off as people in Washington wanting more wine, and ignored it. But the demand went up the chain of command and had to be obeyed. So Goudsmit sent Fischer to southern France to get samples of the wines, the soils, the water, the grapes, etc. It turns out that some regions have naturally radioactive soils. So that shipment answered their questions.

IS THAT AN ATOM SMASHER?

When I knew Howard Fuite in the 1940s, he was a technician in the Physics Dept. of Northwestern Univ. Although he never attended a college, he was trained to set up equipment for students to use in the laboratory sessions of their physics courses. Occasionally he also served as the tour guide for tourists wishing to see the huge Science and Engineering Building, where Howard worked. On one tour, there was a man who was obsessed with wanting to see a particle accelerator, which were in the news after World War II, although Northwestern did not have one. Every time that Howard showed the group a large instrument, the man would say, "Is that an atom smasher?" "No, this is an X-ray diffraction instrument, etc." After 7 or 8 times that Howard heard the question, he was getting peeved. So when they came to a large stack of condensers, Howard said, "And that is an atom smasher." The man said, "Oh, thank you, Dr. Fuite, thank you!"

BOB GEHRZ' EXPENSIVE TAXI RIDE

Bob Gehrz visited for his first time the American Astronomical Society's headquarters on Florida Ave. in Washington, DC. When he was finished, he hailed a taxi to take him to the Hilton Hotel, where he was to stay. He did not realize that the hotel was only a block away. Not only was that an expensive way to go one block, but the taxi fares in Washington were based on zones. By crossing Florida Ave., he crossed into the next zone and the fare was even more than for going one block.

LEO GOLDBERG'S PASSPORT

Leo Goldberg was on his way to Europe where he was to attend the IAU General Assembly as its President. When he changed planes in New York, he was asked for his passport – in the days when they were not checked at the airport of origin. Looking over the counter, he was shocked to see his wife's picture in the passport! He and his wife kept their passports in the same drawer, and he had grabbed the wrong one. He had to stay overnight in New York until his passport was transmitted to him.

MR. GUSTAFSON'S FAMILY

Mr. Gustafson had been a bank vice president in a North Dakota town just south of Canadian border. After retirement he was bored and became the tour guide at the Lick Observatory in California. He told me this story.

Early in World War II before the US became involved, Canada was involved as part of the British Empire, and the Royal Canadian Air Force flew missions over Germany. Their pilots on R&R often went across the American border because they could have more fun in the US before we were involved and didn't have rationing. One day two RCAF pilots were sent to Gus' bank because they had driven their car across the border, got into a wreck, could not drive the car back, and did not have enough money to get it fixed. Gus said that he had ways of transferring money across the border. While that was happening, he asked them where they flew. They said that they flew from England to Berlin, dropped they bombs, and flew back. He asked them if they ever flew over the little town of …… (where his mother and sister lived). They said that it was always on their route. In fact, on their last mission they realized after leaving Berlin that they had one bomb left in their bomb bay. They looked down, saw that village, and dropped their remaining bomb. Gus was immediately sick but tried not to show it. He did not want to tell them they may have killed his mother and sister; it would spoil their R&R. However he worried throughout the war.

After the war he got the first letters from Germany. His mother wrote that they had difficult times during the war, but they survived safely. The Allies dropped only one bomb on their village, but it was a dud.

THE WHITE DWARF SPECTRUM

Daniel L. Harris, III, sat in on W. W. Morgan's class on spectral classification at the Yerkes Observatory. Late in his course, Morgan liked to project on a screen the spectra of unusual stars or sometimes composites, and challenged the students to classify them. Once he projected the spectrum of a white dwarf on the screen. Harris immediately said that it was the star 40 Eridani B. Morgan was dumbfounded at Harris' knowledge of stellar spectra. But Harris explained that it was obviously the spectrum of a white dwarf because of its broad lines and was taken with the Yerkes 40-inch telescope. 40 Eridani B was the only white dwarf bright enough for Morgan to photograph its spectrum.

DANIEL HARRIS AND HIS PARENTS' CAR

Daniel Harris visited his parents in Cleveland in winter. One evening they went to the theatre. While they were inside it snowed and, when they went outside, all the cars were covered with several inches of snow. Dan offered to drive. On the way home he complained to his mother that her car was not driving well. She said that she had just had it tuned. At their home they parked the car in their covered parking garage.

The next day they went to use it car and the snow had melted off of it; it was the wrong color. The key had fit and the model was the same, but they had driven someone else's car home. They called the police and reported a stolen car.

THE SKUNK IN THE OFFICE BUILDING

Art Hoag tells that once someone left open the back door of the office building on Kitt Peak and a skunk walked in. A hasty conference was held – outside – to discuss how to remove him. Someone suggested that since skunks like lettuce, they should drop pieces of lettuce down the corridor to lure him out. So they did. And then another skunk ate his way into the building!

THE SCORPION IN THE SINK

An astronomer developed some plates in the 0.9m telescope darkroom. When she turned on the lights, there was a scorpion in the sink! She, of course, screamed hysterically and reported the event to Art Hoag. So Art simply mounted a hammer above the sink with a sign saying, "Use in case of a scorpion."

TOM HOAG'S STORY FOR HIS FATHER

At Art Hoag's funeral his son Tom told the following story about his father. Not long before Art was in the hospital with tubes in his mouth and nostrils. Tom told him about the man who called 911 and said in a strong dialect,

"Ma wif has died. Kin you pick her up?"
"Yes. Where do you live?"
"Eucalyptus Street."
"How do you spell that?
Long pause. Then,
"If'n I haul her over to Oak Street kin you pick her up there?"

The tubes flew out of Art's mouth and nostrils as he burst out laughing.

PAUL HODGE AND THE COBRA

Paul Hodge was observing with the Schmidt Telescope at the Boyden Observatory on South Africa. To save time on his last night, he put his whole night's plates into the developer and left the darkroom by the double door. When he was about to re-enter the darkroom, a sizable cobra slithered in just ahead of him. What to do? Proceed with the processing, risking death with the cobra, or turn on the lights and ruin his night's work? He chose the former. After the plates were safely in the stop bath, he turned on the lights and saw the cobra coiled up beside the pipes under the sink.

HUBBLE'S COMET

W. W. Morgan told me that Edwin Hubble's first position was as a young astronomer at the Yerkes Observatory. He used the 24-inch reflector to take photographs of star fields. Early one night he photographed what was later called R Monocerotis. It looked like a comet, so Hubble walked down the hill to the train station in Williams Bay to send a telegram to Harvard, announcing his discovery.

Then he continued to observe it and saw that it showed no motion, as a comet would. He had made a serious mistake at the beginning of his career! (It was later named Hubble's Variable Nebula.) So he spent the rest of the night looking for a comet – and he found one! So at dawn he walked down the hill to the train station and sent another telegram to Harvard, giving a corrected position.

BOB KRAFT'S WINE TASTING CONTEST

Once a year Bob Kraft had friends over to his house for wine tasting. One event was a contest to see who could judge some wines the best. Several wines were served and the guests had to evaluate them on a scale from 1 to 100. Doug Lin was a frequent guest at these occasions. He admits to not being a connoisseur of wines, but he reasoned as follows. He did not think that Bob would insult his guests by serving any wine less than an 82, and he could not afford to serve a wine greater than 92, so he voted 87 each time. He usually won.

GERRY KRON AND HORACE BABCOCK'S CHRISTMAS PRESENTS

Gerry Kron and Horace Babcock exchanged Christmas presents each year. The unwritten rule was that they should be completely useless tools. One year one gave the other a tape measure in which all the numbers and lines had been deleted. Another year one gave the other a screw driver with a shaft that rotated freely within the handle.

KUIPER'S NEW CAR

Gerard Kuiper bought a new large Oldsmobile in the 1950s. Actually, he could not resist a bargain by buying a demonstrator model.

Unfortunately it was colored pink. His wife Sarah loved it, but Gerard did not think that it was dignified enough, so he had it painted gray when he could spare it. Of course they did not paint the interior of the trunk, so when the huge trunk door was opened, the gray car with the pink trunk interior looked like a yawning hippopotamus.

THE SUICIDE RATE AMONG LABORATORY SPECTROSCOPISTS

Barry Lutz was a laboratory spectroscopist at the Northern Arizona University. We once discussed why the suicide rate was so low among laboratory spectroscopists. He said that it is because it is hard to jump out of a basement window!

ROGER LYNDS' BACKPACK

Roger Lynds is physically fit and energetic, well into his 70s. He often went on strenuous hikes with Bill Livingston and others. But on one long difficult hike in hot weather Roger was lagging behind and the others were silently wondering if Roger was beginning to show his age. When they reached their destination, hot, sweaty, thirsty, and tired, Roger reached into his pack and pulled out a huge watermelon!

THE NOISY ELEVATOR

Martin McCarthy, S. J., often worked at the Vatican Observatory at Castel Gandolfo, south of Rome. That building includes the Pope's summer residence. On the first floor are his working rooms, on the second his bedroom, on the third the Observatory offices, and on the roof the Schmidt telescope. There was a rustic elevator going to the roof that operated by occupants pulling on ropes. One cloudy night Martin and a colleague, being bored, had fun by going up and down the elevator, making a fair amount of noise in the process. Once as they passed the second floor the door burst open and Pope John XXIII, in full girth and stature, cried out, "Jesus Christ, what's going on here?" (Martin explained to me that in Italian the exclamation "Jesus Christ" is not considered to be swearing.) Martin replied, "We're just having some fun, Father." The Pope replied, "Well, please be a little more quiet," and slammed the door.

MEINEL'S DUCKS

Marjorie Meinel bought some young ducks with the thought that they would make good meals after they were fattened up. But the ducks became pets of the children. One day she roasted and served one. Once the kids heard what is being served, they said "Which….. one….. is …….it?" Marjorie said "It's George". The kids said "No, not George!", burst into tears, and would not eat it.

MILLIKAN'S MAID

Physicist and Nobelist Robert A. Millikan once heard his black maid answer his home telephone. She said, "Doctor Millikan's residence. Yes, sah. Yes, sah. No, sah, he is not the kind of doctor that does anyone any good." Millikan loved to tell that story.

W. W. MORGAN'S SPECTROGRAPH

Each Saturday morning the Yerkes Observatory had an open house and let the public in to see the motions of the huge 40-inch Refractor. The faculty had to take turns with that chore. Morgan hated that because he was uncomfortable before large crowds, which sometimes numbered over 100 people. One Saturday the telescope had on it the spectrograph with which Morgan had done most of the work in his career. When he showed the crowd how the telescope moved in two coordinates, the dome rotated, and the observing floor moved up and down, he accidentally ran the floor into the spectrograph. On the one hand his heart almost stopped, worrying whether he damaged his spectrograph and on the other hand he was embarrassed at having made a stupid mistake in front of a large crowd. He tried to ignore the accident but a little boy piped up, "Hey, mister, didn't that hurt the telescope?"

TRYING TO STUMP MORGAN

W. W. Morgan was the world's expert in classifying spectra. In fact, he developed the system of two-dimensional classification (temperature and luminosity). He taught his skill to many students. Once some students decided to stump him, so they superimposed the spectrum of Uranus onto that of a supernova. They gave it to Morgan to classify. Morgan said, "I've never seen anything like this in my life. It looks like a supernova superimposed onto a spectrum of Uranus."

PETER PESCH'S LUNCHTIME ACQUAINTANCE

Peter Pesch, who is shorter and thinner than average, had lunch at a counter and happened to sit next to an obese man. The man looked at Peter in admiration and asked "What do you do for a living?" Peter said "I am an astronomer." The obese man said, "Well, I buy and sell scrap metal. There is no use spending much time looking at scrap metal – scrap is scrap. So all my negotiations are ones over meals. And I cannot eat a grilled cheese sandwich while encouraging a potential customer to eat well. So I eat three big meals daily, and that is why I am so heavy.

KEVIN PRENDERGAST'S BREAKFAST

Kevin Prendergast told us one day that he had had a heavy breakfast: TWO cups of coffee and TWO cigarettes.

THE COWBOY IN PARIS

William "Bill" Reid obtained a Ph. D. under Chandrasekhar and then joined the faculty in applied math at the University of Chicago. On a sabbatical in Paris, he occasionally attended an American cowboy movie. In one the hombre came into the bar, pounded his fist on the counter, and demanded "gimme three fingers of red eye". The French caption read "Un martini, Monsieur!"

DOUG RICHSTONE'S FIRST WORDS

Doug Richstone did not talk until the age of three. During that year he was at a picnic and his first words were "May I have another slice of watermelon?" When asked why he did not talk earlier, he said that people gave him everything that he wanted so he never had to ask.

ELIZABETH ROEMER'S MEASURING ENGINE

Elizabeth (Pat) Roemer was very particular regarding the accuracy of any screw-type measuring engine that she used to measure astrometric positions of comets on photographic plates. She had a favorite engine that she brought from the US Naval Observatory when she moved to the University of Arizona. Once when she was discussing an inferior screw on another measuring engine, she said with a twinkle in her eye, "You have to go to the Navy to get a good screw."

THE SECRET PLANS FOR THE NIGHT

Paul Routly was a graduate student working under Lyman Spitzer at Princeton on a Ph.D. thesis. The two came to Mt. Wilson to obtain the observations. However, student Routly was not allowed to stay in the Monastery, so he stayed at the nearby Mt. Wilson Hotel. It was an old wooden building with paper-thin walls. In the early evening Lyman Spitzer came to Paul's room to make plans for the night. Their conversation ran something like, "Let's do that job after the moon sets and it will be dark." A neighboring guest heard that and called the Manager that there were some criminals in the next room.

C. D. SHANE'S TIP TO THE PORTER

C. D. Shane took a trip by train across the country in a sleeper. At the end he felt that he should tip the porter, but did not know how much. So he asked him. The porter said that the average tip was $5, so he gave him a $5 bill. The porter said, "Thank ye, Sir. Not many people give as much as the average."

MICHAEL SNOWDEN'S BUMPER STICKER

Michael Snowden was a strong advocate of the metric system. He could not understand why the U.S., like most of the rest of the world, didn't convert to that logical system. He had a bumper sticker on his wall from the U. S. Metric Association that said "Go Metric". Art Young came in once, looked at the sign, grabbed a ruler, and found that it measured exactly 3 x 10 inches in size!

Art did not stick around for an explanation. Michael had been arguing for years with the U. S. Metric Association to actually go metric. He told them that they used 8.5x11-inch paper for their correspondence and newsletters, and English units in all their work, like their bumper stickers. It is only recently (due to Michael?) that they have actually converted to the metric system. Strange.

THE ABT BEER

Lewis E. Snyder was a radio astronomer at the Univ. of Illinois. He spent several years at the Max Planck Institute in Munich. He told me that about once a month he drove to Brussels for a stock of Abt beer. It was made in a monastery north of Brussels and sold only in the pub across the street. It is known for its high alcoholic content (6%?). Once on a trip to Brussels for a conference, I took Daniel along. He took the streetcar to the pub and bought a dozen bottles of Abt beer. The box provided would only hold 11 bottles, so Daniel said "I will drink the last bottle here", to which the clerk said "Monsieur, this is not Budweiser!"

JOEL STEBBINS' FRIENDS

Joel Stebbins once said to his wife, "The Jones are back from Europe. Let's go and visit them before they get their Kodachromes back."

JOEL STEBBINS' MEDAL

Joel Stebbins was active in astronomy for many decades. Also as Secretary of the American Astronomical Society for many decades, he figured that he had attended more meetings of the Society than anyone else. He thought that he deserved a medal for that, but despaired that anyone else would think of awarding him a medal. So he decided to give himself a medal. At one meeting he pulled out of his pocket a large shiny silver disk and gave it to himself. Then he read the back of it that said, "Hershey's Chocolate."

BENGT STROMGREN'S CAR

One evening after dinner at the Van Biesbroeck's we young faculty members walked to the Yerkes Observatory for our evening's work. In front of the Observatory on the circular drive was Bengt Stromgren's car with the lights on, the motor running, and the driver's door open. But inside the building there was no light in his office and he did not answer to a knock. So we walked to his house, where he was, and asked him about his car. You could see the lights flashing in his brain! He had been to the campus in Chicago, stopped at the Observatory to check his mail, and, by habit, he walked home.

HAROLD UREY'S LECTURE

I heard this second- or third-hand, so I cannot vouch for its accuracy, but I love this story.

Harold Urey was asked to give a lecture in a number of mid-western cities, fairly close together. It seemed reasonable to provide him with a car and chauffer, rather than schedule many airline flights. During the trip the chauffer expressed admiration to Urey for his being able to give that lecture. Urey said "Nonsense. You have heard it so many times and you could give it yourself. I'll tell you what. In the next city they do not know me from Adam, so let's change clothes and you give my lecture". That was done, and the lecture went well. But after many lectures there is a person in the audience who tries to ask the most difficult question that he can. After a man did that, the chauffer said, "Really, Sir, that is a very simple question. In fact, I will let my chauffer answer it!"

THE CONFERENCE FOR GEORGES VAN BIESBROECK

Georges Van Biebroeck put on his office door a poster for a "Conference Dedicated to the Late Georges Van Biesbroeck". He wrote on top of it "Not yet!"

THE YONI PAPER

Gerard Wasserburg, a geochemist at Caltech, had a childish sense of humor, like insisting that his group that studied lunar rocks and related science be called "The Lunatic Asylum" in published papers. Editor Chandrasekhar did not like such childish humor in the ApJ, and would not let him use that name in the Journal. So Wasserburg decided to get back at him.

Wasserburg and others wrote a paper (ApJ, 157, L91, 1969) which obviously concerned "Continuous Uniform Nucleosynthesis Theory".

The acronym for that is not something that you would use in the ApJ, and it never appeared in the paper. But Wasserburg translated it into Hindi as "Yoni", which only Chandra would understand. Knowing that Chandrasekhar read all the papers that he published (that could be done when the ApJ published 150 pages every six weeks), Wasserburg did not use that word in his paper. But he slipped it into a figure caption, hoping that Chandra would not bother to read all the figure captions. Chandra did not catch it and it was published, much to Chandra's chagrin.

THE WILLIAMS BOYS

Near the Yerkes Observatory in Williams Bay, Wisconsin, lived a family named Williams (no connection with the name of the village) that had two teen-age boys. I took them for rides on my sailboat on Lake Geneva, as I did for anyone else I knew and who would enjoy sailing. About 50 years later I received a phone call from a man who identified himself as one of the Williams boys. I agreed to meet him at a motel north of Tucson, where he was attending a meeting. There I met a tall stocky man in flowing robes who is now called Lama Kunga Gyaltsen, who is the Spiritual Director of the Sakya Tsarpa Rime Dharma Center in Boulder and is the founder of the Colorado Dharma Coalition during half years; during the other half years he is at the Kunga Tenpheil Ling organization for SAMSARA Mediation Ensembles in Kathmandu, Nepal. He is probably at the Terger Osel Ling Monastery outside of Kathmandu.

OLIN WILSON'S PIPE

Olin Wilson smoked a pipe. He kept the pipe, loose coins, and his keys in his left coat pocket. In the right pocket was the packet of tobacco and the kitchen matches that he used to light the pipe.

One day he was engaged in a conversation in the corridor of the Mt. Wilson Observatory on Santa Barbara Street. He automatically extracted the pipe with his left hand, the tobacco and a match with his right hand, lighted his pipe, and tossed the match on the floor. But the pipe didn't catch. So he used another match, but the pipe didn't light. After a while he realized that there were burned out matches all around him on the floor and his pipe still was not burning. So he knocked out the tobacco into his hand and out fell a penny.

JOHN WOOD'S SWIMMING SUIT

John Wood once had to go in November from Indiana University to Long Beach, California, to visit a contractor. On arrival at the motel, he quickly discovered three things: (1) it was pleasantly warm that November evening, (2) the motel had a large heated swimming pool with no one in it, and (3) he had not packed a swimming suit. But the temptation was too great, so he decided to sneak to the pool in his boxer shorts. So he grabbed a towel and just after he closed his bedroom door, he thought, "Whoops, the key."

So John went to the front desk and asked the clerk for a spare key to his room. The clerk started to search around under the counter. John tried to wrap the towel around his waist, but it wouldn't reach. He drummed his fingers on the counter to show his impatience. Just then there was a loud applause in the next room, the banquet was over, and about 150 people streamed into the lobby to see John standing there in his underwear.

THE MESSY DARKROOM

One summer while I was observing occasionally on Mt. Wilson, I happened to talk with a new janitor. He said that he had been a commercial photographer all of his career, but when he retired, he found that he did not have quite enough money to live on. So he worked on

Mt. Wilson as a janitor during the summers when they needed some extra help. We talked about various things, and then he said, "Some of the astronomers sure make a mess in the darkrooms. Why, the other afternoon I had just cleaned up the 100-inch darkroom when the astronomer came in. He was not in there very long, and then came out. I looked in to be sure that everything was spick and span, but found developer on the ceiling, fixer on the floor, and a huge mess. When I came out he was still there. I know I shouldn't have, but I could not resist saying, 'Excuse me, Sir, but is it necessary to make such a mess in the darkroom?' He looked me straight into my eyes and said, 'Have you ever tried working in complete darkness?'"

I thought that was a funny story and told it at the dinner table. At the head of the table was Sir Richard Woolley, the Astronomer Royal, who was on the 100-inch. I was on the 60-inch, and according to the mountain protocol, sat next to the head. During my story he interjected questions like "He was a commercial photographer?" And at the end of the story he said, "And that was me."

ZWICKY'S MYSTERIOUS STUDENT

Fritz Zwicky was an arrogant person. He openly showed his distain of other astronomers ("Those stupid bastards"), which was justified because he had some breakthrough ideas and the evidence to back them that were ignored for too many years. He was one of the great astronomers of the 20th century. But he grated nearly everyone. He taught a course on mechanics in the Physics Dept. of Caltech, which was one of the most difficult courses in the school. But some faculty and grad students in Physics decided to play a trick on him to "take him down a peg." So they enrolled a fictitious student in his class. The student turned in perfect homework papers, with the help of the faculty, but never showed up for classes. Zwicky became more and more curious about this remarkable student and wanted to meet him. He knew that the student would have to show up for the final exam.

Exams at Caltech are given on the honor system. So shortly after the beginning of the exam a student, ostensibly to go to the men's room, delivered a copy of the exam to the faculty and grad students waiting in a nearby room. They worked through the problems, but the handwriting became worse and worse. Finally it ended in a scrawl saying "I'm too drunk to go on" and Zwicky never met his remarkable student.

VOLUME INDEX

Aarseth, Seth L. 163
Abt, Helmut A. 1 4 6 7 9 10 12 15 16 17 18 20 24-28 30 33-41 49-70 122-133 138 159-166 202-208
Abt, Karl A. (father) 10 11 12
Abt, Karl W. Abt (brother) 12 70
Abt, Margaret S. Abt (mother) 11 12 70
Adams, Walter 208
Aitkin, Robert G. 50
Aller, Lawrence 48
Al-Naimiy, Hamid M. K. 172
ALSOS 13
American Astronomical Society 123-126 131
Angermeyer, Gusch 97-98
Ap stars, magnetic stars 58-59 157-158 192
Argue, Noel 208
Arp, Halton C. 43 209
Astrophysical Journal 20 43 65 68 102-115 130 174 192
AURA (Association of Univ. for Research in Astron., Inc.) 42-44
Baade, Walter 5 209
Babcock, Horace 219
Bai, Li 143-144 185
Beijing 139
Ball, Lucille 11
Balmorhea State Park 31
Batten, Allan 51 68

Battle of the Bulge 12 13
Big Sur 50
Biggs, Eleanor S. 102 120
Bjornarud, Chris 209-210
Blanco, Victor 48
blue stragglers 157-158
Boesgaard, Ann, Hans 61
Bolenbaugh, Andrew (son) 15 22
Bolenbaugh, John (father) 15 16
Bolenbaugh, Virginia. (mother) 15 16
Bolenbaugh Richard (son) 15 17 22 72
Boonyarak, Chayan 177
Boyce, Peter B. 111
Brandt, John C. 6 28
Brunetti, Ivan 210
Camino del Diablo 1
Caltech 14 23-24 173
Campbell, Bruce 117
Canyonlands 72-73
Casparis, John O. 33-38
Cerro-Tololo Inter-American Observatory 41 46 61
Chaffee, Fred 72
Chamberland, J. W. 27
Chandrasekhar, Subrahmanyan 102 113 210-211
Charles Darwin Research Station 93
China 68 134-156 193
Chinese discoveries 151-156
Chinese National Observatory 148
Chinese observations 4
Chinese opera 147 186
Chinese stamps 147-148
Chupp, Edward L. 109
circumcised, circumcision 16 17 69

Clarke, Arthur 211
Clark, D. H. 4
Claspy, W. P. 8
colorblind 77
Conaway, Daniel V. (son) 10 68 225
Corbally, Christopher 21
Cowley, Anne 65 68 138 212
Crater Elegante 1 2
Crab Nebula 3
Crab Nebula supernova 4 6 8
Dalgarno, Alex 103 107
Deutsch, A. N. 4
Deutsch, Armin 212-213
Devine, Steve 95-96
Docobo, Jose 167
Dos Passos, John 143
double stars 57 116-117 163-164 192
Duncan, John C. 4
Duyvendak, A. 4
Easter Island 80-88 187
Eastern Schism 8
Edmondson, Frank 213
Eggen, Olin 25 213
Englert, Fr. Sabastian 80-81
European observations 8
exoplanets 117-119
Fan, Charles 27
Ferrell, Richard 18
Field, Charles 89-99
Fiji 76
Fisher, Russell A. 13 14 214
Fitter, Julian 94
Flagstaff 32

Ford Foundation 42 159-162
Fountain, John W. 6 9 68-70 74 146
French Polynesia 75
Fu, Du 143-144
Galapagos Islands 85-99
Gallagher, John 65
Garmany, Catherine D. 53
Gehry, Frank 147 168-169
Gehrz, Robert 215
geoglyphs 160
Gerbaldi, Michele 175
German soldiers 12
Goldberg, Leo 215
Goudsmit, Samuel 13
Grand Canyon, 22 71-72
Grant, Gordon 48
Gray, Richard 21
Great Wall 139
Greenstein, Jesse L. 14 17 19 25 57
Guinan Edward 65
Gum Nebula 27
Gustafson, "Gus" 215-216
Hale, George Ellery 109 173-174
Harris III, Daniel L. 216-217
Hawaii 61
He, Xiang-Tao 147 186
Healey, Tom 78
Heiser Arnold 28
Henyey, Louis G. 23
Hiltner, W. Albert 30
Hoag, Arthur A. 217-218
Hodge, Paul 218
Hoehn, Iona 112

Hoffleit, Dorrit 121
Honeycutt, Robert 65
Hong Kong 138 146
Hoyle, Fred J. 5
Huang, Emperor Qin Shi 134-135 142
Huangshan (Yellow Mt.) 134
Huang, Yao 148
Hubble, Edwin 218
Hucke, Alicia, Eugenio 81-88
hyperfine structure 22
IAU General Assembly 68-70 174-175
Irwin, John B. 32
Ishida, Goro 65-67 69-70
Istomin, E. 11
Izaemon XVII 66
Jackson, Robert H. 11
Jackson State College 55-6
jade 146 186
Jamestown, NY 11
Japan 61-70
Japanese observations 4
Jeffers, Hamilton 50
Jeffers, Robinson 50
Ji, Zhang 144 185
Johansen g. 11
John Muir Trail 71
Jordan 170-172
Joy, Alfred H. 20 47
Jugaku, Jun 62 65 67-69
Kabuki plays 65-66
Kamb, Barclay 1
Kennocutt, Robert C. 112
Kent State Univ. 55

King, Robert 18
Kinkaku-ji (Golden Pavilion) 62-63 69
Kitt Peak National Observatory 27 28 33-35 37-46 49 122
Koch, Robert 65 138
Kondo, Yoji 65 138
Kovtyukh, V. V. et al. 23
Krogdahl, Wasley S. 14
Kron, Gerald 219
Kroupa, Pavil 163
Kuiper, Gerard P. 14 48 159 219-220
Kunming 134 139
Kyoto 62-63 68-70
Lallement, Rosine 165
Lampland, C. O. 4
LAMOST Observatory 27 140
Land of Standing Rocks 72
Latham, David W. 163
Lavdovsky, V. V. 4
Leonard, Frederick 1 2
Leung, Kam-Ching 65 68 138 175
Levato, Hugo 52
Levy, Saul G. 53
Lhasa 148-150
Lick Observatory 25
Lin, Doug 138 219
Local Interstellar Bubble 165-166
Lundmark, Knut 4
Lynds, Roger 43 220
Maglev train 144-145
Mahler, Gustav 182
Malaysia 178
Marsh, Olin 11
Marshall Space Flight Center 79

Martin, J. C. 8
Mayall, Nickalus U. 4 211-212
Mayor, Michel 117
Marmilliod, Jean-Claude 157
McAlister, Harold 65
McCarthy, S. J., Martin 220-221
McDonald Observatory 27 29 48
McMath, Robert R. 33
Meinel, Aden B. 27 33 38-41 140
Mendoza, Eugenio V. 28
metallic-line stars (Am) 57-59 157 192
metal-rich stars 163-164
Meyer, D. 6
Michaud, George 58
Miczaika, Gerhard 41
Michelson, Albert A. 19
Micronesia 77
Mihalas, Dimitri 103
Miller, William C. 1 4 5 6 22-24
Millikan, Robert A. 221
Minkowski, Rudolph 32
Misawa, Paul 63
Mishima, Yukio 63
Moore, Elliot 28
Morgan, William W. 27 49 52 218 221-222
Moseley, Peter 72
Mulloy, William 82
Mount Wilson Observatory 19 23 120-121 173
Mt. Fuji 62
Multi-Mirror Telescope 27
Munch, Guido 17 19
Museum of Northern Arizona 5 6
Nan Modal 77

Nanjing 140
Naresuan University 179
Navajo Canyon 6 7 9
Needham, Joseph 134 151-156
Noh plays 66
NOIRLab 44
Northwestern Univ. 12 13
Oak Park River Forest High School 12
Ogdon, John 11
Ohio State Univ. 61
Oort, J. H. 4 19
Optics Center, UA 27
Osawa, Kiyoteru 62
Osmer, Patrick 61-3 72
Osterbrock, Donald 103
Parker, Eugene 18
Pauling, Linus 1
Pecker, Jean-Claude 69
Perry, Ruth 112
Pesch, Peter 28 79 222
Petri, Egen 11
Peterson, R. T. 11
Petra 170-172
petroglyphs, 6 7 73
Philip, Prince 96-97
Physical Review 13
Picasso, Pablo 167
Pinacate National Park 1
Ponape 76-77
Popper, Daniel 65
Potola 149
Pourbaix, Dimitri 51
Poveda, Arcadio 65

Prendergast, Kevin 222
publication studies 122-133 192
Purple Mountain Observatory 140
quasars 43
Quitobaquito 1 2
Rashid, Hussein 170
Ready, Fr. David 87
Reid, William 71 223
Roberts, Morton S. 17
Rock Pile 31
Roemer, Elisabeth 25 223
Roosevelt, Franklin D. 11
RV Tauri stars 26
ryokan 68-70
Sagan, Carl 28
Saito, Kiyoshi 64
Santiago de Compostella 167-168
Sandage, Allan R. 16 17 18 19 23 50 71
Sanford, Roscoe F. 23
Sandberg, Carl 143
Schulz, Charles 184
Schwarzschild, Martin 19
Sexton, Janice 112
Seyfert, Carl 32
Shane, Charles D. 25 224
Sharp, R. 1
Sherrill, Robert 111
shock waves 23
Shore, Steven 68
Shostakovich, Dimitri 182
Simkin, Tom 99
Sirindhorn, Princess 176 179
shunks 206 217

slavery 148
Slettebak, Arne 57 158
Snowden, Michael 224
Soonthorthum, Boonraksar 179
Smith Clayton 28
Spain 167-169
Stebbins, Joel 29 225
stellar interiors 58-59 157-158
Stephenson, F. R. 4
Strom, Stephen E. 103
Stromberg, Gustav 47
Stromgren, Bengt 19 225
Struve, Otto 14 47 113
supergiants 30
Sykes, John B. 211
Tahiti 75-76
Tamm, Ronald 65 138
Temple, Robert 151-156
Thailand 175-180
Thompson, Harold J. 38
Tibet 134 148-150
Tohono O'odham Nation 39-40
Trimble, Virginia 4
Truk Lagoon 77-78
Trumpler, Robert J. 52
University of Chicago Press 104 174
Urey, Harold 226
Van Biesbroeck, George 30 100-101 140 205 226
Van den Bos, Willem H. 50
Visiting Professor Program 54-56
Warner, Brian 69
Wasserburg, Gerard 226-227
Webbink, Ronald 65 138

White Mesa 4 5 74
Wild, Earl 11
Williamson, R. A. 6
Wilson, Albert G. 17 19
Wilson, Olin C. 19-21 227-228
Wilson, Ralph P. 120
Wilson, Robert 65 138
Whitman, Walt 143
Wolff, Sidney 68
Woolley, Sir Richard 229
Wood, H. John 228
Woods, David 64
Worley, Charles 50
W Virginis 23
Xi'an 142-144 185
Xinlong Observatory 140
Yerkes Observatory 14 28 173
Zhang, Yuzhe 140
Zhao, Gang 148
Zion National Park 73-74
Zwicky, Fritz 229-230